Praise for *Lov*

"*Love Your Mother* is a beautiful ... to action for climate justice from diverse voices—a must-read!"
—**Leah Thomas**, author of *The Intersectional Environmentalist: How to Dismantle Systems of Oppression to Protect People + Planet*

"Wrap yourself in this book like a lovingly made quilt. In bringing together these stories, *Love Your Mother* shows how many beautiful, powerful ways there are to do this work, and in doing so, it issues a warm invitation: *Join us*."
—**Katharine Wilkinson,** coeditor of *All We Can Save: Truth, Courage, and Solutions for the Climate Crisis*

"This book is a mighty collection, a great read for anyone who cares deeply to care about Earth, community, and the climate crisis. Dr. McDuff offers up bite-size stories of inspiring climate action from across all fifty states, from a spectacularly diverse and accomplished group of women."
—**Leah Stokes,** author of *Short Circuiting Policy*, and Anton Vonk Associate Professor of Environmental Politics, University of California, Santa Barbara

"Women have been at the forefront of the climate battle from the start, and this book is proof of it. If we have a fighting chance of coming through these decades, it's because of them!"
—**Bill McKibben,** author of *The End of Nature*

"Mallory McDuff does a beautiful job of taking us along the journey of women tackling the everyday task of living while protecting the planet and their communities. Whether mom, madre, mama, or mother-identifying person, all can find themselves within the words of each story."
—**Heather McTeer Toney**, vice president of community engagement, Environmental Defense Fund

"Expressed in these fifty stories is a wild love for Mother Earth and her children—a love for all of us, alive together, indivisible. These fierce American voices filled me with two emotions I had not allowed myself to experience in a long time: pride and hope."

—**Will Harlan,** author of *Untamed: The Wildest Woman in America and the Fight for Cumberland Island*

"Through vivid, thoughtful storytelling, McDuff's profiles emphasize a timely truth: climate leadership isn't a monolith. Matriarchs, farmers, writers, rebels, scientists, doctors, innovators, influencers, teachers—all of us, in short—have a home in this movement, if we choose to seek it."

—**Georgia Wright,** co-creator of the podcast *Inherited*

"*Love Your Mother* is a collection of unadulterated and unconditional love stories—for our fellow humans, for our children, for our mothers—and for Mother Earth. And who doesn't want to read a love story . . . or fifty!"

—**Jill Drzewiecki**, Gender-Responsive Education Specialist, Jesuit Refugee Services

"Mallory McDuff's heartfelt portrayal of these inspiring women demonstrates the power of individuals to address the most consequential issue of our time—the climate crisis. Women, especially those from frontline communities, will bear this burden far more than others."

—**Stephen Mills,** vice president, Strategic Partnerships and International Relations, The Climate Reality Project

"I couldn't put it down. Each story is filled with the joys and struggles of women who are fully engaged in the climate movement. Surely their tenacity and resilience will spark further change across the country. I am deeply grateful for this book of hope for our times!"

—**Mary Evelyn Tucker,** cofounder,
Yale Forum on Religion and Ecology

"Yes, the women profiled here are investors, teachers, scientists, and farmers; but they are also mothers, daughters, volleyball players, cooks, and sculptors—and in sharing their stories, Mallory McDuff offers fifty examples of what we can do."

—**Peter Turchi**, author of *Maps of the Imagination:
The Writer as Cartographer*

"Mallory McDuff offers a tapestry of beautiful stories about the kind of transformative love and momentum needed to usher in climate justice. These women in different communities and walks of life, understand the stakes of this moment because their daily lives are materially shaped by the climate crisis."

—**Aby Sène-Harper,** assistant professor and researcher in conservation social science, Clemson University

"From India's Bishnois women—the original tree huggers—to courageous Catholic sisters and more, women have played a vital leadership role in efforts to protect the planet. Mallory McDuff tells these stories with her characteristic verve and depth of belief."

—**The Rev. Fletcher Harper,**
executive director, GreenFaith

LOVE YOUR MOTHER

LOVE YOUR MOTHER

50 States, 50 Stories, and 50 Women United for Climate Justice

Mallory McDuff

Broadleaf Books
Minneapolis

LOVE YOUR MOTHER
50 States, 50 Stories, and 50 Women United for Climate Justice

Cover image: Getty Images
Cover design: 1517 Media

Print ISBN: 978-1-5064-6444-2
eBook ISBN: 978-1-5064-6445-9

For Jill and Brian,
Love and laughter, friends forever

AUTHOR'S NOTE

Climate storytelling feels like a lifeline and a way forward for me as I navigate the world with my daughters and my students. In these pages, I hope you'll find points of connection as well. For the majority of the stories in this book, I conducted interviews by phone or Zoom and used reliable secondary sources as background material. The women reviewed their essays for accuracy and edited for clarity. In some cases I was unable to speak with the person featured, and so I relied solely on secondary sources. Any factual errors that remain are my own. For ease of reading, I chose not to use endnotes and listed all sources for each section in the bibliography.

For structure, I organized the states according to regions defined by the US Census Bureau (although looking at a map, I still can't believe Oklahoma is in the South). I also recognize the limitations of a colonialist concept like "states," but chose that framework as change often begins close to the ground at state and local levels. Given discussions about Washington, DC, as a possible candidate for statehood, I featured one woman, Rhiana Gunn-Wright, who described both Illinois and DC as home. The scope of this book does not include the US territories, sites for innovative climate organizing as well.

It's worth noting that these stories feature a moment in time: since I completed the research and writing, several women in the book have moved; others have finished school, found new jobs, and given birth to babies. Court cases described in the stories are ongoing. Both the positive impacts and the inherent shortcomings of historic federal climate legislation are unfolding. We are a fluid country and world—capable of imagination and transformation—in so many ways. That is one reason why we must share our climate stories.

To care about climate change, you only need to be one thing, and that's a person living on planet Earth who wants a better future. Chances are, you're already that person—and so is everyone else you know.

—*Katharine Hayhoe*

Joy is the justice we give ourselves.

—*J. Drew Lanham*

STATES AND STORIES BY REGION

(WEST)

Alaska	Nevada
Arizona	New Mexico
California	Oregon
Colorado	Utah
Hawai'i	Washington
Idaho	Wyoming
Montana	

MIDWEST

Illinois	Kansas	Nebraska
Indiana	Michigan	North Dakota
Iowa	Minnesota	Ohio
	Missouri	South Dakota
		Wisconsin

NORTHEAST

Connecticut

Maine

Massachusetts

New Hampshire

New Jersey

New York

Pennsylvania

Rhode Island

Vermont

SOUTH

Alabama	Georgia	Mississippi	Tennessee
Arkansas	Kentucky	North Carolina	Texas
Delaware	Louisiana	Oklahoma	Virginia
Florida	Maryland	South Carolina	West Virginia

CONTENTS

CONTENTS

CONTENTS

INTRODUCTION

LOVE ABOVE ALL

It was below freezing on a bitter winter morning when my friend Jill and I bundled up our young daughters—ranging in age from four to eleven—in puffy coats and warm hats and drove an hour away to protest the proposed construction of a new coal-fired power plant in the mountains of Western North Carolina. We'd packed the requisite PB&Js and water bottles for the three kids, dark coffee in insulated mugs for us. When we arrived at the site, our girls stared longingly at the counter-protestors, who'd harnessed more resources than our rag-tag group.

"Why do they get Chick-fil-a?" my fifth grader asked. "I think they have hot chocolate too!"

"Go grab a sign," I said, dodging her question. "Take your sister and Aja with you."

Our daughters soon joined the twenty or so adults chanting, "No more coal! No more coal!"

I remember wondering if the revolution should require better snacks.

It must have been 2009 or 2010—when we took our children to prayer vigils and marches before they had a stake in their weekend plans. But that morning as we huddled together for warmth, Jill turned to me with a grin, pointing toward the two younger girls.

"Do you hear what I hear?"

I stepped close enough to listen to them, two high-pitched preschool cheerleaders: "No more COLD! No more COLD! No more COLD!"

They weren't aiming for irony, just reading the chilly vibe of the day.

All we could do was double over in laughter—at our good intentions, the levity of children, the power of persistence, even when we didn't know how the story would end. We were showing up the only way we knew how—as stressed-out mothers, good-hearted educators, and glass-half-full community members—sharing our fear for the future and also our delight in the day.

I couldn't raise my children, now twenty-three and sixteen, in a climate crisis, without the women I hold close to my heart. My treasure chest is these friendships, as well as the long-distance mentors whose work I read and Instagram posts I follow, although they do not know my name.

It was my dear friend Jill, actually, who proposed the idea for this book, which stemmed from her job as a gender-responsive education specialist for Jesuit Refugee Services in Rome, Italy, far from her home base in Appleton, Wisconsin. In refugee camps across the world, she's seen the disproportionate impact of the climate crisis on girls and women, who make up 80 percent of those displaced by climate change. She lives the words of diplomat and climate leader Christiana Figueres: "Educating young women and empowering women to come to decision-making tables is the strongest thing we can do for the climate."

Women in countries like the Central African Republic, where I served in the Peace Corps in my twenties, will bear the brunt of impacts from emissions of countries like the United States. In this context, I wanted to illuminate collaborative climate leadership by women that might be our exit ticket and opportunity through this crisis. As Drs. Ayana Elizabeth

Johnson and Katharine Wilkinson remind us in their anthology *All We Can Save,* "The climate crisis is not gender neutral."

I've spent my adult life with young people—some paralyzed by climate grief and anxiety—and others who have harnessed uncertainty into action and reflection in their communities. Among them is one of my former students, Kelsey Juliana, the lead plaintiff in a court case suing the federal government, demanding a climate plan to protect the constitutional right of youth to a healthy life. The case asserts the US government knew about the impacts of burning fossil fuels more than fifty years ago, but continued to incentivize putting carbon into the air. We know this much is true.

This book is not only about prominent leaders like Kelsey calling for structural change toward climate justice. It's also about women of diverse ages and backgrounds in every state in the country, from Alaska to Arkansas. Growing up in Alabama and raising my daughters in North Carolina, I've yearned for relatable stories from other places—towns and communities that aren't mentioned in the lists of Top Ten Climate Activists of any given year. Most people have heard of the epic work of Greta Thunberg, but her message is actually about the collective that involves so many others, not the lone individual making headlines.

As Rebecca Solnit writes, "The qualities that matter in saving a valley or changing the world are mostly not physical courage and violent clashes, but the ability to coordinate and inspire and connect with lots of other people and create stories about what could be and how we get there." It's harder to tell the narrative of hundreds of people coming together around a common goal as opposed to a single story in a spotlight, but those are the stories we desperately need to hear and to tell. When she was in middle school, my youngest daughter, Annie Sky, called out my own inclination to "hero worship" when she asked, tongue in cheek, "Do you like Greta more than me?"

It's actually for the love of my children and students that I offer these climate stories of women working together across the heartland and the coasts, from rural regions to urban centers. Half of the fifty women featured here are BIPOC (Black, Indigenous, and People of Color), with six of them being Indigenous. To select the women from each state, I looked for a diversity of stories about changemakers on multiple scales: what unfolded before me was the intersectional reality that climate justice is racial justice. Valuing some lives over others, which is inherent in white supremacy, only stokes the climate crisis. Dr. Rhiana Gunn-Wright, who helped to develop the Green New Deal, reminds us that climate justice is about access to a healthy environment for all. The concept of climate justice includes the built environment and is a result of economic and social systems. "It's about changing power relationships," she says.

The stories you'll find feature activists, sculptors, poets, climate scientists, farmers, journalists, oceanographers, students, teachers, and more. This book presents a sampling of ways to love the earth, not a comprehensive overview. The women represent a spectrum of identities and backgrounds: queer and straight; cis-gendered and nonbinary; atheist, agnostic, Jewish, Muslim, Buddhist, Christian, and more. They range in age from fourteen to seventy, although I didn't include their current ages to account for the inevitable passage of time, and I integrated their chosen pronoun into the text.

Organized by region, the stories guide the reader in three sections: the intersection of their lives, background, and momentum. Each woman's story thus starts in the middle, looks back to the past, and then forward to the future.

For every person in the book, I sought to respect their identity as they claim it, and I bring my own identity and lens as well: I'm writing as a white, straight, single mother with a strong extended family, a teacher of environmental education, a person of faith, raised in the Deep South. With my daughters, I live on a college campus in a 900-square-foot rental with an expansive view of the Appalachian Mountains, a herd of cattle, and a donkey named Tallulah.

🍃 🍂 🍁

During one month of research for this book, my brother's home in Alabama was pummeled by hurricanes, and my sister and her family in Washington State were trapped inside their house due to smoke from wildfires. At a neighborhood potluck that same summer, I shared updates from *Love Your Mother* with a friend who teaches landscape architecture at the University of Georgia. "What do you want to happen after people read your book?" he asked me.

In response, I described how many of the interviews I'd conducted transformed into authentic friendships, collaborations, relationships.

"I hope readers connect with others around climate justice by using their unique strengths," I said. "If they make art, create art together. If they want to march, then find their people and take to the streets. I want us to build friendships to power through this mess in an innovative way together."

That's how I personally responded during the research, making connections without even trying. When I talked with physician Dr. Suzanne Bartlett Hackenmiller from Iowa, we stayed on the phone for two hours, discussing her research about nature therapy, but also sharing our experiences as mothers of teens. With Sarah Bellos from Tennessee, a CEO revolutionizing fashion and fabric, I told stories about students from my college weaving wool from our sheep through the fiber arts program.

This relational angle to climate justice is accessible for all of us in our increasingly nonbinary world, however we identify. So I want to introduce you to fifty of my friends—some are old friends, most are new. These women now populate my social media feeds and inspire what I bring to the classroom and how I advocate with my elected officials.

Scientists say individual actions can only account for 30 to 40 percent of the change needed to confront this "code red for humanity," as the UN Secretary General calls our climate crisis. In contrast the US military is the institution that consumes the most fossil fuels in the world. But as activist Xiye Bastida writes in *All We Can Save*, "A vibrant, fair and regenerative future is possible, not when thousands of people do climate justice perfectly, but when millions of people do the best they can." We have seen collective public pressure achieve what once seemed impossible, including the historic passage of federal climate legislation in Congress in 2022.

In every story I heard, the call to love was epic: love for the Earth, children, community, elders, land, oceans, wildlife, home. Love was the force. Not some syrupy sentiment needle-pointed into a pillow but the deep, unrelenting love of a mother who knows fracking is poisoning her children—and she seeks another way. This love felt unstoppable, even in the face of intentional public misinformation by corporations harming the earth in the name of protecting financial gains.

Climate writer Mary Annaïse Heglar described this best as: "Wild love. This love is not a noun, she is an action verb. She can shoot stars into the sky. She can spark a movement. She can sustain a revolution."

Women have been on the forefront of climate science as early as 1856 with the discoveries of Eunice Newton Foote, a physicist and inventor from Seneca Falls, New York. Indeed, she discovered the greenhouse effect three years before Irish-English scientist John Tyndall got public credit. Because she lacked access to education and training, others considered her an "amateur" although she wrote a scientific paper documenting the results of her experiments. As Katharine Wilkinson writes, "What might Foote have achieved if she had access to Tyndall's training and resources?" (As an aside, Tyndall was accidentally killed by his wife who gave him an overdose of the wrong medicine.)

It's now time to learn from and celebrate the story of Eunice Newton Foote and so many women charting the way in each state in this country and across the world. For example, I want my students to understand the climate leadership of trans women, such as Precious Brady-Davis, with the Sierra Club and author of the book *I Have Always Been Me*, as well as Izzy Rose Laderman, who advocates from the intersection of climate and disability justice.

As my own research expanded, I started referring to the women by their home state, so in conversation with friends, I'd mention: "Oh, I'm writing about Ms. Oregon today!" or "Gotta run now, I'm interviewing Ms. Maine!" At first, I kept this inside joke to myself, sensitive to the gendered vernacular of beauty pageants. (Full disclosure: I was voted "Spirit of Junior Miss" in 1984 in Alabama.) But once I shared this habit with my interviewees, the parlance took on a new power. Dr. Lou Weber, who teaches ecology in Indiana, said her extended family planned to make her a "Ms. Indiana Climate Change" sash with photovoltaic cells at their next reunion.

"I'm honored to wear the sash," she said. "Maybe you're reclaiming a tradition for a new era."

The need for structural climate action—not just climate change but "everything change," as Margaret Atwood says—feels like heady stuff when my ordinary life includes teaching classes, parenting daughters, cooking dinner, paying bills. Late at night I'm often content to zone out on Instagram until it's time to turn out the light. But when I'm mindlessly scrolling, I see these new friends inviting me into a movement bigger than this household of three girls, my precious but small world. In my daily life and in a climate crisis, I want to be with people who aren't giving up, especially when I'm overwhelmed and pissed off by the lack of political will to protect the common good by those in power.

As audio producer Georgia Wright said in the climate newsletter *Hot Take*, "Don't let fear eat you alive. You're already here, aren't you? Instead of panicking, stay your course . . . We're in this bitch for the long haul."

❀ ❀ ❀

The second summer of my research, my friend Jill and her daughter returned to Western North Carolina after a two-year absence. As we talked at a mile-a-minute pace, Jill told me she hoped these climate stories reminded women not to underestimate their scrappy efforts at change.

"When I'm working globally, I see so many women caught up in the day-to-day," she said. "And we don't get the chance to zoom out and see the potential outcomes of our collective efforts."

As teenagers now taller than their mothers, our daughters, Annie Sky and Aja, walked ahead to the Swannanoa River on campus. They lay on their backs on a wide, flat rock and watched the water flowing past them, as it had when they were toddlers. I pretended not to stare at their long limbs and sculpted cheekbones, but in that moment, I saw one truth: I don't know how this story will end, but I am not alone. I am a part of a larger climate movement that could turn this monumental threat into a collective call for justice based on unstoppable love above all.

SOUTH

ALABAMA

MEDIA MATTERS:
HOLLYWOOD GETS A LESSON ON CLIMATE CHANGE

ANNA JANE JOYNER

CLIMATE COMMUNICATIONS EXPERT

Perdido Beach, Alabama

"When Rick Joyner tells you to evacuate, do it," Anna Jane Joyner said as she recalled her father's advice during the 2020 hurricane season at her family's compound on the Alabama Gulf Coast. In the night she and her husband followed her dad's counsel and fled to higher ground. The water flooded neighborhoods and shredded rooftops, leaving coastal communities without electricity for two weeks.

Anna Jane is no stranger to the impacts of climate change. She's harnessed the power of storytelling while fighting coal-fired power plants and advocating for Hollywood screenwriters to integrate climate into their scripts.

"My family has lived on this land for over a century," she told me, "And I might be the last generation to experience this beauty." Sea levels along this coast are expected to rise a range of eighteen inches to four feet in the next century, with lives, property, and ecosystems at stake.

She lives beside the Gulf Coast waters where I grew up swimming as a child, so her stories strike home for me. "From my front yard, I can see pelicans, gulls, and dolphins, but I also experience so much trauma and anxiety around hurricanes," she said. "Hurricane Sally ravaged the land. This place looked like a war zone. It's like we need a new vocabulary for talking about hurricanes."

Now she's bringing that climate reality to TV shows and movies in Hollywood, but with a different lens from apocalyptic science fiction. She knows that four in ten Americans experienced a climate disaster in 2021 alone. And her real-life narrative is rooted ten feet above sea level in one of the most climate-vulnerable places in the country. When Anna Jane couldn't find climate storylines on the big screen—with relevant dramatic tension and complex characters—she looked closer to home, seeing her own family story as one shared by many.

> "It feels strange that we've gone through the worst fire season and hurricane season, but none of that complex reality is reflected in our film and TV shows. We need intimate, human stories about the climate because that's what we're experiencing—that's the scale."

She discovered her life's work during a study abroad program in college, which challenged her conservative upbringing and connected her to diverse voices caring for the Earth.

Anna Jane and her dad, Rick Joyner, weathered deep-seeded disagreement about climate, a rift rooted in politics in this country. Her prominent, socially conservative, evangelical father—a pastor with followers in the millions—cut off her college tuition when she majored in communication rhetoric with a minor in Environmental Studies and devoted her life to climate communications. For six months they didn't speak, but their love prevailed, much like in the movies.

In the Showtime series *Years of Living Dangerously*, Anna Jane told her story: one episode chronicled her unsuccessful efforts to convince her father about the importance of climate science. At the time, she was working on the Asheville Beyond

ALABAMA

Coal Campaign with the Creation Care Alliance, which lever-
ages the power of faith communities for the climate. The documentary featured her outdoor wedding as Anna Jane stood in
a locally-designed gown and recited her vows: "I promise to
honor God, nature, and our families," she said, lifting up values
she learned from her parents.

She's also shared climate stories in her podcast "No Place
Like Home" with cohost Mary Anne Hitt. "We're two Southern women, climate activists, and friends," said Anna Jane,
"and the stories explore the spiritual, personal, cultural, and
emotional dimensions of the climate crisis." Her work reveals
how climate affects everyday lives, including her decision not
to bring children into a warming world.

Anna Jane believes in the need for relatable and accurate climate
storytelling as only 2.8% of scripted TV shows or films from
2016 to 2020 mentioned climate change or related keywords like
sea level rise. The organization she founded called Good Energy
aims to increase that number to 50 percent by 2025. With that
goal in mind, she launched the online resource *Good Energy: A
Playbook for Screenwriting in the Age of Climate Change* with support
from Bloomberg Philanthropies and actors like Mark Ruffalo.

"It feels strange that we've gone through the worst fire
season and hurricane season, but none of that complex reality
is reflected in our film and TV shows," she said. "We need
intimate, human stories about the climate because that's what
we're experiencing—that's the scale."

This initiative began after she reviewed a script for the CBS
show *Madam Secretary*, which featured an evangelical daughter
in conflict with her father about climate change. More than
five million people watched the episode. From the deck of her
family compound, next to her aunt and grandfather's house on
the bay, Anna Jane knows the real-life drama of living with a
climate crisis. And she wants to expand the reach of these true
stories to the stars of Hollywood and beyond.

ARKANSAS

THE SUPERPOWERS OF SOIL:
HOW REGENERATIVE AGRICULTURE CAN SAVE THE EARTH

DONNA KILPATRICK

RANCH MANAGER AND LAND STEWARD

Perryville, Arkansas

Grazing cattle isn't just about moving a herd from one pasture to another. It's also about the regeneration of life from the soil.

"In my job, I have the privilege of making a positive impact on the ecosystem by the holistic management and movement of cattle," said Donna Kilpatrick. She's a longtime rancher and land steward whose use of holistic planned grazing promotes healthy ecosystems to support biodiversity and grow healthy food.

"In the early morning, I have the opportunity to observe what's going on—watch the tree swallows, evaluate forage growth, and look for dung beetle activity," she said.

The 1,200 acres she manages in Arkansas belong to the nonprofit Heifer USA, a program of Heifer International, a global development organization that works with communities to end hunger and poverty while caring for the earth. The site also includes a three-acre organic vegetable garden. Many people know about Heifer International from their program that donates animals to farmers in other countries, but Donna was hired to scale up agricultural production and training for farmers in the United States, especially in the Delta region of

Arkansas. Among the things she's especially proud of is the all-female ranching team she works with each day.

When she began this job, the soil quality was poor since most of the land had been leased to individual farmers for thirty years without much oversight. As part of her plan for the soil, the ranch went from 25 cows to a herd of 300 cattle, 35,000 chickens, 400 pigs, and 200 sheep.

"We're showing farmers how to restore ecosystem health but also providing online markets so they can sell at fair prices."

"Animals can be key to diversifying the forage, as well as restoring and aerating the soil," she explained. "We're showing farmers how to restore ecosystem health but also providing online markets so they can sell at fair prices."

Regenerative agriculture is a farming method that goes beyond sustainable agriculture to nourish the earth. Some studies suggest we have less than 60 more years of quality soils to grow our food, she said, so her passion for farming comes with a sense of urgency to renew the soil. Poor soil quality can result from a multitude of factors, including soil erosion, overgrazing, monocultures, tillage, and more.

In the late 1980s, Donna discovered this love for farming while studying at Warren Wilson College, where I also teach and live. As we talked on the phone, a student mowed the pasture in front of my house, so I understood how working on the farm could change the trajectory of someone's life. Growing up about an hour away in Brevard, North Carolina, Donna discovered this campus as a place of acceptance for a gay woman in the South.

"I grew up with the message that being gay meant I was doomed for the fiery furnace," she said. "I had a very loving and supportive family, but I didn't find my people until I started working on the farm at Warren Wilson."

It was during college that she discovered the magical world of getting to work early at daybreak. These days, she wakes up at 4 a.m. to drink her coffee, watch the sunrise, and catch up on email before heading to the pastures.

"I had an awakening in college," she said. "Agriculture drew me to the soil, to the animals, and provided a sense of place. I found my roots. My North Star is regenerating the land here and training farmers how to regenerate theirs." We all need a North Star, she believes, to guide decision-making in our personal lives and our work.

Working on a dairy farm in Western North Carolina and as a Peace Corps volunteer in Ecuador gave her more experience before joining the staff at Heifer International. And among her contributions in the past four years has been her work with the Grassroots Farmers' Cooperative, an online market that ensures farmers have stable markets for their products. She's also using the ranch as a hub to teach holistic grazing and land management through the Savory Institute, a global network for regenerative farming.

Both the climate crisis and pandemic amplified the need for localized food production for the future. For Donna that meant she and her staff ramped up their online instruction, which provided useful teaching tools even after farmers returned to hands-on learning at the ranch. In videos she stands as a solid and confident presence, calmly explaining concepts, wearing her wide-brimmed hat and Carhartts, clearly rooted in the land and healthy ecosystems for all.

And what's Donna's advice for the interns she oversees? Find a North Star and work where you can be a steward of the planet, and find ways to care for the land.

DELAWARE

NAMETAGS, ROUND TABLES, QUESTIONS:
A RECIPE FOR A CLIMATE CONVERSATION

LISA LOCKE

DIRECTOR OF PROGRAMS

Rehoboth Beach, Delaware

The young people brought their questions, and they weren't afraid to ask. "How can older generations support younger generations who will live with the effects of climate change for most of their lives?" "How can we promote climate science and prevent the spread of misinformation?"

While some families avoid this kind of questioning at intergenerational events, fearing answers can turn political or divisive, these eighty people came together in Delaware to talk and listen to each other at what they called a Climate Conversation. Each round table hosted a deliberate mix of three groups: high school and college students, faith leaders from diverse traditions, and senior citizens.

With notebook in hand, Lisa Locke walked between tables, checking on each group in her role as director of programs for Delaware Interfaith Power & Light (IPL), an organization with affiliates in forty states and a mission to engage diverse religious communities in confronting climate change. "There were some big lessons from this first event," Lisa told me. "First, the room was abuzz, and I was struck by how engaged people were in the process. And second, people said they usually didn't have the chance to talk about climate with others." One of her student interns, Kirit Minhas, said the climate was a source of constant anxiety for him, but the dialogue provided a point of connection.

The effects of climate change in the state include eroded beaches, coastal flooding, damaged farmlands, and increased salt levels in estuaries and aquifers, among others. For the past eight years, Lisa has spearheaded Delaware Interfaith Power & Light, known by its acronym DeIPL. In one action their members joined with advocacy organizations such as Oceana and Citizens Climate Lobby to link hands at Rehoboth Beach, facing the water as a statement of their faith and opposition to offshore oil drilling. Several years ago, I'd traveled across the country to write about climate action by congregations, so I was especially keen to hear about this work.

After the first climate dialogue in 2018, Lisa submitted a proposal to Energize Delaware, which has funded two similar Climate Conversations every month since. Part of the inspiration for the project, Lisa said, came from the work of scientist Dr. Katharine Hayhoe, author of the book *Saving Us: A Climate Scientist's Case for Hope and Healing in a Divided World.*

"She says the most important thing you can do to address climate change is to talk about it," Lisa said. "It's easy to think that talk is cheap, but it's an important place to start."

Now other Climate Conversations begin with prompts such as: What comes to mind when you think about climate change? Have you experienced the impacts? What capacity do we have as individuals, communities, states, and nations to affect the outcomes?

"It's so common to feel paralyzed about these big issues, but when you talk about how we can work more effectively together, there's a shift," she said. "Climate change is a moral issue, and that's one place to find common ground."

Delaware IPL has hosted more than thirty-five conversations involving more than 1,000 people. While one conversation won't change the world, participants from twenty events responded to surveys assessing the impacts of the dialogue. Afterwards, 39 percent of the 300 respondents reported being more convinced of the threat of climate change; 54 percent felt more hopeful, and 66 percent felt more inspired to take action.

"I don't want to exaggerate the meaning of these numbers," Lisa said, "But when you talk to each other, it can make a difference."

In Grand Rapids, Michigan, Lisa spent fifty years as a member of a nondenominational church with a long history of open and respectful dialogue. She took what she gained from those discussions and her time working as an administrator for the church when she transitioned to environmental work and moved to Delaware. Not only did she learn generative and respectful conversation from her faith community but she also credits her parents for shaping her ethos, with their own dedication to civic responsibility and delight in nature.

The work of Delaware Interfaith Power & Light brings together communication as well as action. Their Faith Efficiencies program coordinates energy audits, remediation, and education for faith communities and their facilities, including tracking energy usage and savings for a year. Across the country, IPL affiliates engage in similar initiatives: in Washington State, they've helped to stop coal export terminals, a case that went to the US Supreme Court, and in North Carolina, the organization helps congregations adopt renewable energy initiatives like solar panels.

"It's so easy to feel paralyzed about these big issues, but when you talk about how we can work more effectively together, there's a shift."

It might start with a conversation around a card table, but identifying shared values is a way that everyone, regardless of age, background, or religious beliefs, can stand together for the good of the world.

FLORIDA

CLIMATE AND EDUCATION:
LIGHTING A FIRE FOR CHANGE

CAROLINE LEWIS

EDUCATOR

Miami, Florida

"So I want to be an arsonist," Caroline Lewis explained to the young people and their parents, who exchanged quick glances with each other. "I want to light fires in elected officials, frontline communities, and more."

Using her "teacher voice" at this film screening, Caroline held everyone's attention, even more so when she spoke passionately about climate change with both directness and respect.

"If you understand the climate science, then that fire should be lit," she said, helping young people cultivate that same passion in south Florida, where king tides cause residents in some areas to don rainboots on flooded streets as water rises through the porous limestone bedrock from the saturated ground.

As one of seven children, she was born and raised in Port of Spain, Trinidad and Tobago, where she played outside from sunrise to sunset with cousins and neighborhood friends. In south Florida Caroline has brought insights from her twenty-five years as a teacher and principal—from Port of Spain to Miami—to integrate science and climate education. The climate crisis is a threat multiplier, she understands, especially for those living at or below the poverty level in the Miami area, where flooding has increased 400 percent in the past ten years. The state of Florida is most at risk of flooding from sea-level rise of all fifty states, with predictions that the ocean could rise three to four feet by 2100.

That's why in 2010, she launched the CLEO Institute—Climate Leadership (through) Engagement Opportunities—because she saw a critical need in the community. The 2.7 million people living in Miami-Dade County were at the center of the climate crisis, but there was little community education focused on causes, impacts, or solutions. The purpose of the nonprofit is to work with educators, elected officials, students, and frontline community partners to build climate literacy and support for bold action, something Caroline knows how to inspire. The amplifying impacts of a warming world extend beyond rising seas and include food, water, rising temperatures, and health vulnerability. In Miami, climate accelerates gentrification when those with financial resources buy property on higher ground and price out those who've spent their lives there.

From hosting climate town halls in the Little Haiti neighborhood to supporting "Climate and Me" summer camps for youth, Caroline aims to meet people with concerns related to their home turf—and figure out together how climate intersects with their needs and values. So when she spoke at the screening of "From Pittsburgh to Paris," a documentary featuring her story, one student asked a question, and Caroline encouraged him to speak up so everyone could hear.

"Loudly!" she insisted.

"Why did we withdraw from the Paris Climate Accord?" he asked, repeating his question in a clear voice the second time around.

"The President withdrew," Caroline replied. "But the rest of us didn't."

Another student asked for help identifying next steps to try and fix the climate crisis.

"That is such a good question, and I think you should run for office, young man," she said, jumping in to share the importance of voting and running for elected office as strategies to impact climate policy.

As she explained the three climate powers each person holds, the crowd listened, fully engaged.

"You have the superpowers of your voice, your vote, and your purse."

Offering the example of her own family, where talking about climate shifted the conversation among those with different political beliefs, she explained, "I have very strong Republican stock in my family." While they may not agree on politics, she said her family members have grown to support her work on behalf of the climate because they care about her.

> **"You have the superpowers of your voice, your vote, and your purse."**

As a teacher myself, I sometimes wonder if one educator can truly make a difference, especially at the end of a semester, when energies run low but climate challenges have amplified. Through Caroline's work, I see the multiplier effect of education for individuals, communities, and institutions.

"I'm a science teacher, so when I see a problem, I want to fix it," she said. "I have two daughters. I don't want them growing up in a world without the beauty and biodiversity that we enjoy." Not an educator to miss a teachable moment, Caroline repeats a line by William Butler Yeats: *Education is not the filling of a pail but the lighting of a fire.* And she's already lit the spark.

GEORGIA

BEYOND THE STATUS QUO:
BUILDING FEMINIST CLIMATE LEADERS

KATHARINE WILKINSON

CLIMATE AUTHOR, TEACHER, AND STRATEGIST

Atlanta, Georgia

It was a mild spring night in the mountains, and the 500-seat theater was packed, except for the back two rows. I'd come to hear Dr. Katharine Wilkinson in person, after listening to her TED Talk that went viral about empowering women and girls in order to help stop global warming. A group of teenagers tromped into the auditorium with their flannel shirts, Carhartts, and water bottles at the ready. They settled into the back rows, limbs draped across each other, comfortable after three months together at The Outdoor Academy. The students were here because Katharine spent a semester in high school in the same program in Pisgah Forest, North Carolina, a time that influenced her path toward building feminist leadership for the climate.

"There are two powerful phenomena unfolding on Earth," she told us. "The rise of global warming and the rise of women and girls. The link between them is often overlooked, but gender equality is a key answer to our planetary challenge."

Commanding the stage, Katharine shared climate solutions catalogued by the organization Project Drawdown—including drawing down heat-trapping, climate-changing emissions. Among the strategies, she explained, are three areas where securing equal rights and opportunities for women and girls would build resilience and avert emissions: farming, education, and healthcare. The bestseller *Drawdown* was focused on these three areas, and she renewed the vision in the follow-up book, *The Drawdown Review*.

"Women are vital voices and agents for change," she continued. "We're too often ignored or silenced when we speak." As she spoke, I peered toward the back of the theater to see the students sitting up straight in their chairs as she described how women and girls face greater risk of displacement and death from natural disasters.

It's been over two decades since she had her "knife-in-the-heart moment" while hiking with The Outdoor Academy in Pisgah National Forest. There the group came upon an area that had been clear-cut, a dramatic scene that fostered her sense of responsibility in making a difference for the earth. That moment was what led her into student activism and then a doctorate at the University of Oxford, where she was a Rhodes Scholar. Each experience deepened her focus to conserve and connect with the planet.

Katharine reminds us the patriarchal leadership that got us into the climate mess won't get us out of it. "The climate crisis is a leadership crisis," she said. "To transform society, we need transformational leadership . . . girls, women, and nonbinary leaders are showing up as the catalytic change-makers the movement desperately needs."

During an episode of the podcast *No Place like Home*, she described a feminist leadership approach comfortable with both science and stories, art and activism, policy and philosophy, with justice as a through line. "The climate movement has been fractured by battling male egos," she said. "There's a need for more collaborative leadership."

In the bestselling anthology of women writers, poets, and artists, *All We Can Save*, Katharine collaborated with Dr. Ayana

Elizabeth Johnson as coeditors. *Rolling Stone* called the book "a feast of ideas and perspectives, setting a big table for the climate movement, declaring all are welcome." And that's exactly what Katharine aims to do.

She left her role with Project Drawdown in order to build the All We Can Save Project—an organization inspired by the book and focused on nurturing a welcoming, connected climate community, rooted in the work and wisdom of women, with a dose of joy as well.

"For my entire life, every educational institution I've attended and organization I've worked for—they've all been run by white men," she wrote. "I've done liberated work in fits and starts, but never like this. I've never given my days to the work I think the world needs most."

Now in this role, Katharine shepherds a growing set of programs, including All We Can Save Circles, small groups seeding climate dialogue, community, and action. She also co-hosts her own podcast, *A Matter of Degrees*, with policy expert Dr. Leah Stokes, exploring diverse topics such as embedding equal opportunity in a clean-energy economy. But she always brings a levity to the heavy lifting: once after dinner at a conference with Mary Robinson, the former President of Ireland and climate advocate surprised her with a request for some dance music.

"What do you play? Does she foxtrot? Love 80s power ballads?"

"I panicked," she later said, "then went with Motown 'Ain't No Mountain High Enough.'"

"It is a magnificent thing to be alive in a moment that matters so much."

With the mountains, poetry, and a commitment to possibility, she believes "it is a magnificent thing to be alive in a moment that matters so much."

KENTUCKY

FINDING SUSTENANCE:
FOOD JUSTICE AND HEALING TO SUSTAIN
COMMUNITIES AND THE LAND

TIFFANY BELLFIELD-EL-AMIN

FARMER, BIRTH DOULA, RESTAURANT OWNER

Waco, Kentucky

Tiffany Bellfield-El-Amin walks in the footsteps of her ancestors who caught babies, farmed the rich Kentucky soil, and used herbs for healing. "On a given day, I'll visit several pregnant moms and maybe catch a placenta if I assist at a birth," she said. "Then I might meet with farmers through the Community Farm Alliance and later grocery shop for the restaurant."

She works all over her county in rural Kentucky as a farmer, doula, and educator helping other Black farmers grow collective entrepreneurial strength as well as local food. She's steeped in a deep tradition of providing sustenance—from a recipe for making elderberry syrup to strategies for accessing federal agricultural funding.

With dark-rimmed glasses and long dreads pulled into a ponytail, Tiffany lives at the intersection of food, health, and climate justice in eastern Kentucky. She's a third-generation Black farmer in a state with only 600 Black-owned agricultural businesses, or less than 2 percent of all farmers. Today there are only 49,000 Black farmers in the United States—making up only 1.4 percent of all farm owners. This statistic compares to 1920, when nearly a million Black farmers worked the land, representing one-seventh of farm owners, according to *Mother Jones* magazine.

"I grew up farming tobacco," she said, tracing her family's history in this state. "My great-great-grandfather was a slave. My great grandfather was born into slavery and was a sharecropper in Kentucky. My grandfather's family owned land and farmed tobacco. Now I'm part-owner of twenty-six acres, where I grow herbs, edible flowers, and pollinator-specific plants."

After her grandparents died, Tiffany and her mother spent six years trying to secure the land after discovering there was a home equity loan against the house and red tape that threatened to prevent them from paying back the loan. She might have walked away from the land, she said, if it hadn't been for her grandparents' legacy as landowners near Richmond, Kentucky, just outside of Lexington.

But she stayed. Now the farm is named Ballew Estates after her grandmother, and Tiffany transformed the land into a site for permaculture, women and teen retreats, STEM education, and value-added agricultural products, like herbal teas made from elderberries and mint. All of these speak to how the healing of the land connects to caring for herself and others.

Tiffany has seen how better nutrition and health care can promote wellness in times of crisis. "I used to work as a nurse care tech, and most of the health problems I saw could have been prevented," she said. "Healthy food makes a difference in healthy bodies."

And she has firsthand experience in her own life: "I experienced domestic abuse—my leg was broken, I was 320 pounds and diabetic, unhealthy, and not eating right," she told me. "I'd almost lost the land and was mentally exhausted. I went through a big transition in three years. My healing became connected to the work I was doing."

During this time, she turned to her agricultural roots with a holistic lifestyle that included her doula role, her garden, and the land, she said. It took hard, sustained work, but providing sustenance to herself and her community became her guiding principle in her life.

Her next step was to begin work at Community Farm Alliance as a farm-to-table coordinator and later a food justice organizer, providing technical assistance to farmers to ensure they have resources to adapt to the climate crisis with resilient food systems.

"I'm often the only Black farmer and woman at the table in these gatherings," she said. "But I can use my voice to ensure access and justice for all."

In 2020 Tiffany purchased the iconic Alfalfa restaurant with her husband Wali. The couple, married only two years at that point, decided to do the impossible: open a restaurant during a pandemic. But their twist on the dining establishment includes support for Black farmers, a retail space for Kentucky-made food products, and a place to nourish those interested in food systems, justice, and community.

> "I'm often the only Black farmer and woman at the table in these gatherings, but I can use my voice to ensure access and justice for all."

"I wanted a brick-and-mortar place, and I knew I could lean on my farmers to supply local food," she said. Serving Appalachian Southern-style meals, the restaurant is another tool in building community resiliency with a changing climate.

Look at her left arm and you'll see a tapestry of colorful tattoos, including these words—Patience and Pressure—juxtaposed side-by-side. The push and pull together seem to reflect her work in the world. When she needs to ground herself as a mother and a community leader, Tiffany returns to the land where she lives with her husband and daughter in a house built by her grandparents. "My best therapy session is having a full day working and weeding in the garden, using a machete, touching the soil," she said. "I can leave my anxieties behind with my sweat in the ground."

LOUISIANA

WOMEN PUSHING BACK THE TIDE:
BUILDING ALLIANCES AND RESILIENCE
ON THE GULF COAST

COLETTE PICHON BATTLE

ATTORNEY AND CLIMATE ACTIVIST

Slidell, Louisiana

With her feet rooted on the ground and her voice steady, Colette Pichon Battle seems to be the type of person most of us would want nearby in a crisis. When she speaks about climate change displacing millions, she uses measured words to describe strategies to dismantle structural racism, build alliances in community, and provide legal services for equitable disaster recovery. As I listen to her TED Talk, I'm reminded of a cheer from my high school in coastal Alabama: "Rock, rock, rock, rock, steady, eddy, eddy, eddy, rock! Rock steady." So I wasn't surprised to learn she describes her superpower as "seeing patterns in chaos," an apt skill for the organization she founded, the Gulf Coast Center for Law and Policy, now called Taproot Earth.

Working on the frontlines of climate disasters caused by hurricanes, sea level rise, and fossil fuel companies, she knows that recovery for industries has been quick, and recovery for communities has been slow. From Houston, Texas, to Pensacola, Florida, her work brings climate change to the community level, especially with women at the heart of neighborhoods and households.

"We found that the folks most willing to get to know each other were actually women," she said in an interview with *Reimagine*. "When women talk about their communities, it's

"When women talk about their communities, it's sort of like women talking about their children ... So a lot of the moral fabric and the moral movement of a family and of a community is done through the women."

She and her staff used a meeting format called the People's Movement Assembly, which involved Black, Latina, and Asian American women learning about each other's lives and agreeing to reach a vision together. From there, groups of women followed through on actions, such as talking about the climate crisis and extractive industries with elected officials in Louisiana who needed the vote from people of color.

In Bayou Liberty, just north of New Orleans, Colette grew up in the house built by her grandfather, where her mother was born. There, water was a way of life: "The bayou is green and lush and all the things that equal bountiful life," she told TED Radio, "But it is also watery and muddy. You can smell everything."

She remembers the names of particular hurricanes along the Gulf Coast, much as I did growing up in Alabama. During the eye of the storm, family members would get into flat-bottomed boats called pirogues to check on neighbors before retreating to safety inside while the other band of the hurricane passed. But the water became unrecognizable given the severity of Hurricane Katrina. As an adult, Colette practiced law in Washington, DC, but after the destruction of Katrina, she vowed never to leave her beloved Gulf Coast again.

When she first saw the Louisiana flood maps at a community meeting, Colette says her life changed. The maps explained how the thirty-foot surge from Hurricane Katrina could flood her community as well as those in Mississippi and Alabama. She realized the land lost from sea level rise was the buffer to her own home—a buffer predicted to disappear.

"I wasn't alone at the front of the room," she explained. "I was standing there with other members of south Louisiana's communities—Black, Native, poor. We thought we were just bound by temporary disaster recovery, but we found that we were now bound by the impossible task of ensuring that our communities would not be erased by sea level rise due to climate change.

"I just assumed it would always be there. Land, trees, marsh, bayou. I just assumed it would be there as it had been for thousands of years," she said. "I was wrong."

Knowing climate is predicted to displace more than 200 million people by the next century, Colette advocates for preparing for global migration by restructuring social and economic systems rooted in justice, such as investing in public hospitals *before* the impact of climate migration or additional storms like Hurricane Ida. It's not like we don't know what is coming, and Colette knows preparation is a life-and-death matter.

"Climate change is not the problem," she said. "Climate change is the most horrible symptom of an economic system that has been built for a few to extract every precious value out of this planet and its people, from our natural resources to the fruits of our human labor."

What holds clear and steady is her belief of what can be done now. "It's already possible, y'all," she often tells people, with the practical sense of someone who can get things done. Colette knows women who have the most to lose from climate disasters also know what it'll take to plan for the future and anticipate the storm.

MARYLAND
THE IMPACT OF INTERSECTIONS:
CLIMATE CHANGE IS A CIVIL RIGHTS ISSUE

JACQUI PATTERSON

POLICY ANALYST AND ORGANIZER

Owings Mills, Maryland

It wasn't accidental, the way Jacqui Patterson spoke in metaphors about climate justice and civil rights: "If I use my candle to light yours, it doesn't take anything away from my candle," she said. "That's a great analogy for the sheer abundance of the earth."

But rather than distributing wealth as an asset for sharing, the United States has been built on a system that labeled some people winners and others losers. With this inequitable system, some groups—such as communities of color, women, immigrants, children—had to lose in order for others to win, she explained, but it didn't have to be this way.

I listened to her give a virtual keynote in North Carolina as she shared stories of diverse communities and explained the deep roots of structural racism in this country: "It's been more than 400 years since the first person was brought as cargo from Africa," she said, "stripped of generational assets and given to white people as *their* generational wealth."

That systemic enslavement continues today: the net worth of a white family is $171,000, ten times the amount of $17,000 for a Black family, she said. We know that 68 percent of Black families live within thirty miles of a coal-fired power plant, while 70 percent live in counties that don't meet federal air quality standards, perpetuating disproportionate health impacts and higher rates of diseases such as asthma. According to the Center for Disease Control, Black Americans are 40 percent more likely to have asthma than white Americans.

The numbers she presented reflected evidence of climate change as a civil rights issue. This intersectionality—the connections between the climate and race, poverty, health, education, and housing—reflected the heart of Jacqui's work since 2009 as senior director of the Environmental and Climate Justice Program for the NAACP. In her role she turned evidence into action, such as designing a *Coal Blooded Action Toolkit* to help groups like Moms Clean Air Force advocate for climate justice. The step-by-step guide helps communities address the toxins from coal-fired power plants polluting their communities.

But what led her to environmental justice was her work as a Peace Corps volunteer in her father's homeland of Jamaica. In a town outside of Kingston, she witnessed residents suffering from health issues due to particulate pollution and a contaminated water supply from a nearby Shell oil refinery. Their only compensation was the construction of several latrines and a recycling program for the local elementary school.

"People in the community were drinking effluent from the Shell oil refinery," she told me. "I was twenty-three years old at the time, but in retrospect, I want to ask: What were the long-term impacts in this community?"

It was years later when Jacqui learned she'd grown up within ten miles of three coal-fired power plants near her home on the South Side of Chicago. The *Coal Blooded* report gave them all a grade of F, rated for their proximity and harm to low-income communities and communities of color. Before joining the NAACP, she'd also worked in disaster relief after Hurricane Katrina and public health in sub-Saharan Africa.

When the NAACP asked her to build the program on climate justice, she started not by writing policy, but by listening to those affected by the changing climate. She visited communities near the BP oil spill in the Gulf of Mexico and those close to coal-fired power plants. "There is so much power and control wielded by corporations and so few mechanisms for justice," she said. "That put me on a trajectory."

With the goals to reduce emissions, advance energy efficiency and clean energy, and strengthen community resilience, Jacqui's work placed as much emphasis on the assets and opportunities in communities as the vulnerabilities. She collaborated with the 2,200 NAACP chapters across the country to tailor climate justice programs to regional needs. So, for example, in Evansville, Illinois, the NAACP now trains people who've been incarcerated to work in the solar industry.

Jacqui recently started work with a new venture called the Chisholm Legacy Project, a resource hub for building Black frontline climate justice leadership. She told me that as a Black woman, race and gender have been sources of strength and vulnerability for her. "I've been in rooms where despite my experience, people look to the white man for expertise," she said. "The weight of this work is amplified by the emotional pull of deeply caring for the people with whom I'm working, as they are my sisters and brothers."

> **"The weight of this work is amplified by the emotional pull of deeply caring for the people with whom I'm working, as they are my sisters and brothers."**

In conversation, she combines the gravity of her words with levity, ending a story with a light-hearted laugh that makes her feel all the more approachable. "When I first was asked to work on climate change, I thought I'd do it for a year," she said. "But I soon realized climate change isn't just about climate. It intersects with everything. So I'm still here!"

MISSISSIPPI

DO WHAT YOUR MOTHER SAYS:
CLIMATE JUSTICE FROM PARENTS WHO CARE

HEATHER McTEER TONEY

ATTORNEY, WRITER, AND POLITICIAN

Oxford, Mississippi

The cloudless sky was bluebird bright, the February air crisp and cold. As a backdrop, the gleaming-white Capitol building stood behind a banner that read *Fire Drill Fridays*, the weekly climate protests mobilized by activist Jane Fonda and Greenpeace. In 2020 hundreds of people gathered in red coats and hats to sound the alarm for the climate crisis and hear the first Black, first female, and youngest former mayor of Greenville, Mississippi, who worked at the time for Moms Clean Air Force, pushing for climate justice to protect children from harm.

With her red shawl and black knit hat, Heather McTeer Toney took the microphone with the confidence of a politician but the intimacy of a friend telling a story at the dinner table. "I have a question for you all," she asked and paused with a magnetic smile framed by bright-red lipstick. "Have you ever heard your mother say, 'We need to talk!?'"

Nodding and laughing, people in the crowd knew exactly what she was talking about. "You know at that moment, the rest of your life, the trajectory of your very existence hangs in the balance of the words of her decision," she said. "How many of you have ever had that feeling?"

Raising their hands, the audience members were with her, hanging on every word: "Well, take that feeling and multiply it by 1.2 million because that's the number of moms with Clean Air Force . . . Congress, your mothers are telling you, 'We need to talk!'"

"I'm from Mississippi," she continued over the applause, describing how mothers had testified before Congress about climate impacts that ranged from asthma attacks and Lyme disease in children to mercury exposure in the unborn. "And the way we say it is, 'We need to have a come to Jesus meeting!'"

In her former role as national field director with Moms Clean Air Force, Heather called again on the voices of the "moms, dads, abuelas, tías, comadores, aunties, big mommas, and play cousins—we will be there to make sure our children are protected from the impacts of climate change."

This justice work has roots that run deep: her father was a civil rights lawyer, and her mother was a teacher. Raised in Greenville, Mississippi, she attended Spelman College, and then got her law degree from Tulane. But her work as a climate justice leader began at the age of twenty-seven, when she decided to run for mayor and focused attention on the contaminated water in her city.

In 2009 the *Washington Post* ran a front-page story with a picture of a young boy in a bathtub filled with dirty water. The headline above the fold read: "Brown Water in Greenville." Below the fold was a photo of this young mayor.

The story prompted a visit from the first Black administrator of the Environmental Protection Agency, Lisa Jackson, who pulled Heather aside and asked: "You know you're working on environmental justice issues, right?"

But she'd just been trying to get clean water in her community. So when Lisa Jackson asked her to serve as chair of the EPA's local government advisory committee, Heather jumped at the opportunity. Two weeks later, the BP Deepwater

Horizon oil spill happened along the Gulf Coast, affecting her home state of Mississippi, as well as Louisiana and Alabama.

As a mother and a triathlete, Heather often jogged the streets of Greenville and saw firsthand the impacts of climate in the Mississippi Delta, a region rich in culture, but impacted by systemic poverty, oil spills, hurricanes, and flooding. Since 2012 she's lived in the university town of Oxford, Mississippi. Her work in the EPA led to her appointment by President Obama in 2014 as the regional administrator for the Southeast, representing 25 percent of the country's population and the center of environmental justice.

At a Bioneers conference, Heather ended her keynote by projecting a picture of her EPA leadership team with six Black women around a conference table. The photo was taken at the end of the Obama administration before the eighty-five environmental rollbacks of the Trump presidency.

Heather has worked with an army of parents who care about changing perceptions of environmentalists as white liberals. "I am a Black Southern woman from Mississippi—Southern Baptist," she said during the keynote. "I can't go into my church and say we won't have chicken and bacon. It's not gonna work. You can't tell me how to grow food because my people taught yours. But we can have conversations in a language people understand."

"We are going to protect our babies to no end ... and the climate has something to do with all of it."

And parents share a common vocabulary around the health of their children. Heather applies those shared values in her new role as vice president of community engagement for the Environmental Defense Fund.

"We are going to protect our babies to no end . . . and the climate has something to do with all of it."

NORTH CAROLINA

CLIMATE LISTENING:
CREATING FILMS AND CONVERSATIONS FOR CHANGE

DAYNA REGGERO

DOCUMENTARY FILMMAKER

Franklin, North Carolina

Dayna Reggero was riding shotgun. She listened with her whole body, deep brown eyes facing the driver, her head of long, brown curls nodding with compassion. I wondered if she even saw the rural North Carolina road in front of her and the tall pine trees on either side. A mother named Tracey steered the car and shared how she joined with other moms and neighbors—Black and white together—in Stokes County, North Carolina, to fight proposed fracking and the harmful effects of coal ash ponds.

A handicapped parking tag hung from the rearview mirror of Tracey's car. She pulled the vehicle to the side of the road to show Dayna how the coal industry buried coal ash under water. What looked like a peaceful pond in a rural area was actually a death trap. Next, Tracey showed Dayna the location for a test drill for fracking.

"My mother was trying to figure out why everybody was getting sick," Tracey said. "That's how it all began." At age forty-four, Tracey suffered three strokes and a heart attack. It turned out the water in this predominantly Black community was contaminated by heavy metals from the coal ash.

Stories of everyday mothers taking action for the health of their children drove Dayna to produce the five-part film series,

"The Story We Want," a collaboration with Moms Clean Air Force. She traveled to eight states to document conversations about climate and community: mothers fighting sea level rise in Florida, a natural gas blowout in California, and coal ash ponds and fracking in North Carolina.

The first film in the series, "Persistence: The Power of People and Prayer," features Tracey, who lives in Walnut Cove, North Carolina, and Caroline, from the nearby city of Greensboro. These two mothers joined together to vote out politicians who didn't support their demands to stop dumping coal ash—and then started organizing coalitions for the health of their children. The documentary shows Dayna listening intently to their stories, paraphrasing the even bigger narrative of diverse communities working together. She is empathy embodied on the screen and in person.

"These mothers all care about children, their homes, and their community," Dayna told me. "It gives them a connection and a common fight, sometimes even a common enemy. Since filming that story, the court ruled the energy company has to clean up the coal ash ponds, and that's important to remember."

Her work revolves around organizing these conversations, but the film's characters often seem equally spellbound by her grounded yet energetic presence. I've seen it happen myself. People seem to grow taller, stand more upright, when sharing stories with her. Dayna's hand is often on her heart as she listens.

The umbrella organization she founded for her storytelling is called The Climate Listening Project, which aims to collaborate, listen, and then document stories that need to be heard. With that mission she's created films for the Natural Resource Defense Council, the National Audubon Society, and Moms Clean Air Force, then amplifying those narratives through National Geographic, Univision, and PBS, reaching upwards of ten million people.

She's always loved the power of stories. At nineteen she worked in marketing for a local zoo in Florida and started using animals for educational spots on TV about conservation. A decade later, she got behind the camera: Dayna believes her commitment to conservation began as a child when her mother was diagnosed with cancer. "I became a vegetarian and an environmentalist because I didn't want anything to die," she said. "It was like my bargain to keep my mom alive."

> "I became a vegetarian and an environmentalist because I didn't want anything to die. It was like my bargain to keep my mom alive."

She sees herself as more than a filmmaker and environmentalist: her Instagram page includes as many photos of the Great Danes she's fostered as trailers for films she's produced. On social media she describes herself as: *Documentary film director. Impact producer. Listener. Artist. Activist. All curiosity and wonder. In love with forests, in awe of oceans.* And she adds: *I laugh loud.* Her work reminds me of this line from J. Drew Lanham's poem: "Joy is the justice we give ourselves."

In her recent film, "Planet Prescription," she documented the intersectionality of the climate crisis, air pollution, and our health care system. The American Lung Association estimates 150 million people, nearly half of the US population, breathe polluted air that contributes to chronic heart and lung disease. Government subsidies of the fossil fuel industry exacerbate our health challenges, just as they contribute to coal ash pollution. During a virtual panel debuting the film, Dayna shared the news that her own father had died due to COVID-19. The nurses, doctors, and health care professionals who had been interviewed in the film nodded in deep compassion for her. Indeed, Dayna models her belief that listening and sharing can shape our interconnected view of ourselves, our world, and the climate, one story at a time.

OKLAHOMA

THE POWER OF MATRIARCHY:
THE RIGHTS OF NATURE AND MOTHER EARTH

CASEY CAMP-HORINEK

ELDER, MATRIARCH, AND HEREDITARY DRUMKEEPER

Ponco Nation of Oklahoma

In a vibrant yellow traditional dress, Casey Camp-Horinek gazed at a clear glass of water she held in her palms, as if cradling a precious baby. She saw a world where nature—the water flowing in rivers and quenching her thirst—had the right to exist.

"My relative. You are everything," she said from the stage. "You are life itself. You are me, and I am you. We appreciate you beyond words. We ask you to carry our blessings to the Coastal Miwok and the Pomi ancestors, who are sustaining us in this beautiful place."

The crowd at the Bioneers Conference knew Casey as an Emmy award–winning actress, in addition to her starring role as a matriarch advocating for the inherent rights of nature as one strategy to confront the colonizing oil and gas industries on tribal land. As a grandmother and movement-maker, she says humans have strayed from the balance with Mother Earth and Father Sky, and we must find our way back.

"All of our conversations about nature's rights begin around mama's kitchen table," she told the audience. "All my daughters and granddaughters go to MIT . . . " She paused and smiled for effect: "That's Matriarch in Training."

That matriarch training has Casey advocating for communities and countries to create laws that honor the Rights of Nature. This means that ecosystems and natural communities have the right to exist and flourish, and people and governments have the authority to defend those rights. This is a different kind of "law" from the US legal system, which considers nature as property, and owners of that property are given rights to destroy it. Federal environmental laws like the Clean Water Act regulate the amount of damage, rather than prevent it in the first place. The first country to recognize the Rights of Nature was Ecuador in 2008. Two years later, Pittsburgh became the first major city in the United States to do the same.

> "All of our conversations about nature's rights begin around mama's kitchen table."

With Casey's leadership, the Ponca Nation became the first tribe in Oklahoma to adopt the Rights of Nature Statute in 2018. The Tribe also passed a moratorium on fracking, or hydraulic fracturing, a method of extracting natural gas from shale. Fracking on this tribal land from 2009 to 2016 caused 10,000 man-made quakes, a serious liability with pipelines underground. And this land was the site of, as Casey said, "enforced environmental genocide," with landfills leaking methane gas, leaching ponds, asphalt cleaning facilities, and fracking and injection wells.

This legacy of colonization and harm to the land dates back to the forced removal of her family from their ancestral homeland of Nibthaska, now called Nebraska, site of the largest aquifer in North America. "Ni" means water and "Bthaska" means flat or "where the water is at the surface of the earth." As a child, her grandfather walked 677 miles on the Ponca Trail of Tears, and the tribe was forced to cede 200 million acres to retain a small township on what is now the Missouri River. One in three people died during the journey, she said.

Casey's advocacy has also included the push to acquire her Tribe's ancestral territory in what's called the Native Land Reform or Land Back Movement. While her work at home started in the 1970s, she's now focused on global connections for climate justice among Indigenous people.

We know Indigenous people represent 5 percent of the world's population but hold 80 percent of the world's biodiversity. Given that legacy, Casey has studied the Māori people in New Zealand, who recognize the rights of the rivers. For tribes in the United States, these rights provide a possible legal strategy for a better future—in addition to direct action like the thousands gathered at Standing Rock to protest the Dakota Access Pipeline. When Rights of Nature conflict with human rights, the courts weigh in. For all who care about the rights of Mother Earth, she advises us to consider the view of a matriarch:

"Gather your family around your kitchen table. Talk to them about what will happen in the next seven generations. Do you want to breathe? Do you want to eat? If you do, do something. Go to your state government. Go to your local government. Go to your federal government and say: We are part of nature. We want you to enact these laws like in New Zealand with the Wangunui River."

As one of four Indigenous leaders and a matriarch in the documentary *The Condor and the Eagle*, Casey journeyed from the Canadian Tar Sands to the Amazon. Uniting Indigenous climate leaders from the North and South, the film shows their spiritual connections in the midst of beauty and devastation. During the premiere in San Francisco, Casey asked all Indigenous relatives in the audience to come on stage. Smiling and proud, she reminded everyone of the power of Mother Earth: "She belongs to you, and you belong to her."

We're Here. You Just Don't See Us.

FACES OF CHANGE

SOUTH CAROLINA

YOU BELONG HERE TOO:
WRITING AN INCLUSIVE RELATIONSHIP
WITH THE LAND

LATRIA GRAHAM

WRITER AND FARMER

Spartanburg, South Carolina

It was her father's gun, a Ruger P89DC, and he'd taught her to shoot like he taught her to fish, gather crab apples, and soak the watermelon seeds in sugar overnight before pressing them into the red-clay earth of the family farm in South Carolina.

The gun was one of the few possessions her mother kept after his death from kidney cancer. And she wanted Latria to pack it in her bag before a six-week writing residency in the Great Smoky Mountains National Park—in a cabin—by herself—away from cell service.

That spring of 2019, Latria didn't take the gun, but she understood her mom's worries. One year earlier, she'd written an article for *Outside* magazine describing her family's relationship to land they'd owned for more than 100 years. Using survey data, she challenged stereotypes about Black people in the outdoors and also confronted structural barriers affecting people of color. The few national parks in the South, for example, are at least a day's drive away for most people. The title of the essay said it all: *We're here. You just don't see us.*

In response, readers flooded her email, Twitter, Instagram, and Facebook. One described checking into a campground with Confederate flags hanging in the office and asked for recommendations of safe places to camp with her child. Others thanked her for being the first Black writer featured in a magazine more known for buff white rock climbers on the cover.

At the time, she couldn't respond to the hundreds of messages. But during the pandemic and protests for racial justice, she wrote a sequel to that essay, a deeply personal letter to her readers entitled: *Out here, no one can hear you scream.* She described visiting the national park archives during her residency to see what information they had on Black people in the most visited park in the country.

"I left with one piece of paper—a slave schedule that listed the age, sex, and race ("black" or "mulatto") of bodies held in captivity. There were no names. There were no pictures," she wrote.

That essay, with its soft vulnerability and raw candor, changed her life. She'd freelanced since graduate school—writing about NASCAR, mental health, farming, football, outdoor recreation, and more. But these days, she fields requests from *The Atlantic* and the *New York Times Magazine*, as she works on a book about her family's past relationship to the land and the present lives of four Black farmers in the South.

Yet her words couldn't save the family farm in Silverstreet, South Carolina. There as a child, Latria played with cousins in the woods and worked at her father's produce stand. "We were doing farm-to-table before it was trendy and before people won James Beard awards for it," she said.

When her dad died in 2013, she discovered he hadn't left a will. Her half-brother insisted she and her brother buy out his share or sell the property. As a fifth-generation farmer, Latria didn't have the funds to take out a loan, even though she'd continued to pay taxes on the property, one parcel among several owned by family members.

She was paid for her second *Outside* essay the day the farm sold, the anniversary of her dad's death. Her book proposal sold soon afterwards.

"I finally had enough cash to get a loan," she told me. "But it was too late. The land had sold. All I could think of was that saying, 'Someday this pain will be useful to you.'"

In her county, there is only one Black farmer at the Spartanburg Farmer's Market, the same market that denied her father a place to sell vegetables years ago.

"My grandmother had a third-grade education. Everyone in my family has broken their body on this land," she said.

"The rural South is being bulldozed for development. With climate change, the South is flooding, the West is burning. But I have my voice and will use it."

Latria has gotten death threats since she started writing about race, all while tackling diverse issues like the dearth of plus-sized technical gear for outdoor adventurers. "I'm a big Black girl with purple hair," she said.

"I know you're scared. Do it anyway."

"And I'm writing about access to the outdoors for all." On her left arm, she showed me a tattoo with one line from her writing: "I know you're scared. Do it anyway."

When I asked how she manages the real fears of our time, she held up a small plastic timer and demonstrated how she gives herself fifteen minutes to feel overwhelmed or sad—or to focus on a task at work. She's been honest in her talks with young people about the serious role of depression in her life.

"What I'm feeling today, I might not feel tomorrow," she said. "Our bodies keep the score, but we can wake up and try again."

TENNESSEE

COLOR THE WORLD:
REGENERATIVE, PLANT-BASED DYES TRANSFORM THE FASHION INDUSTRY

SARAH BELLOS

CEO

Springfield, Tennessee

Dressed in jeans and an indigo-blue T-shirt, Sarah Bellos stood outside a barn wall decorated with old license plates, a Falls City Beer aluminum sign, and a weathered advertisement for Tabasco. In front of her, a table held Mason jars, goggles and a mask, cotton fabric, and the result of more than a decade of research and development: a natural, plant-based indigo dye.

Her goal that day was to teach fiber artists at home how to dye fabrics using bio-based indigo dye, a powder she'd extracted from the leaves of the plant. But the product she'd developed for her female-owned company, Stony Creek Colors, has transformed the fashion industry with contracts from Cone Denim, Levi's, Wrangler, and Patagonia.

Sarah's life operates on multiple levels: she owns the first US company to grow indigo at a scale viable for the commercial textile industry, but she works in local economies, partnering with tobacco farmers in the Southeast who grow the indigo plants. She's the CEO of her own business, featured by Forbes and Martha Stewart Living. And she's also a mother to two young sons in a small town outside Nashville, Tennessee.

Before I spoke with Sarah, I'd never thought about the dyes in my faded jeans, which were made from petroleum-based synthetic indigo. The fiber artist at the college where I teach described Sarah's work to me as "revolutionizing indigo production for the sustainable textile industry in the US."

When she managed a student farm in college, Sarah didn't see how that experience would translate into the fashion industry. After a stint in Washington, DC as an analyst for corporate social responsibility, she moved to Nashville to start a business with her two sisters, who'd grown up exploring the woods and water around Long Island, New York.

The only catch was that they didn't know what their new business would be. In 2006 they combined their joint expertise and started Artisan Natural Dyeworks, a small textile dye house offering garment and piece dying with plant-based colors, rather than synthetic dyes.

"We were screen printing T-shirts and selling them at farmer's markets and festivals," she said. "It was a hard business model. We wanted to sell plant-based dyes and know where the materials were coming from. But you could pay a lot online for natural dyes like indigo, and there was a 90 percent chance they were actually synthetic."

She discovered there was no scalable source for natural dyes. It was challenging to find verifiable plant-based indigo dyes at the volume and consistency needed for the denim market. So she took a risk and ventured out on her own, focusing on the supply chain.

"I knew the market would soon figure out this was a huge opportunity. It wouldn't be the old way—our country had relied on enslaved labor to produce natural indigo in our past and on toxic chemicals to produce the synthetic indigo dye in our present."

"You would talk to people in denim, and they wanted natural indigo," she said. "I knew the market would soon figure out this was a huge opportunity. It wouldn't be the old way—our country had relied on enslaved labor to produce natural indigo in our past and on toxic chemicals to produce synthetic indigo dye in our present."

Tobacco farmers looking to transition to other rotational crops were an essential part of her business model to grow the indigo plants. Other key partners for Stony Creek Colors, which started in 2012, were chemists, denim mills, and brand names like Patagonia who buy the plant-based dyed denim from the mill.

Sarah told me that the global fashion industry is responsible for 1.7 billion tons of carbon dioxide emissions each year. To produce the synthetic indigo dye used in denim, dye manufacturers combine chemicals like the carcinogenic petrochemical aniline with cyanide and formaldehyde and reach extreme temperatures to synthesize the indigo in a carbon-intensive process.

With the Stony Creek regenerative model, however, they developed a specific method to mechanically harvest the leaves, extract the indigo dye, and return the stem and roots to the soil, storing carbon underground. Preliminary data suggest over 3.8 pounds of carbon dioxide are captured and stored to make the Stony Creek indigo used in a single pair of jeans.

"Both the fashion and agriculture industry are realizing the opportunity to sequester carbon in the soil and improve the productivity of the soil," she said.

To scale up, the company secured $9 million in investor funding in 2021. Sarah's team provides the seeds or seedlings, while the twenty farmers they work with plant and maintain the crop. Six years later, the bet is paying off. Sarah "kept the pedal to the floor" despite crises such as a change in the seed supply source the same year she had her first child.

"I would tell other women to keep your focus and don't scale back on your goals. You can work incredibly hard at any scale, selling natural-dyed shirts at a music festival or selling natural dyes to Levi's. We are investing in innovation across the entire agricultural value chain."

TEXAS

SHARED VALUES:
CONNECTING THE CLIMATE
TO WHAT WE CARE ABOUT

KATHARINE HAYHOE

CLIMATE SCIENTIST

Lubbock, Texas

"Do you believe in climate change?" a man once asked Dr. Katharine Hayhoe at a barbecue restaurant in Lubbock, Texas. After attending the church service where her husband is a pastor, he approached their table to chat and soon learned she was a climate scientist.

"No, I don't believe in climate change!" she responded adamantly—a response that took him by surprise. She went on to explain that climate change isn't a belief system but a phenomenon grounded in science. We know the Earth's climate is changing due to observations and data, she said, and because real people are affected, especially the poor. It's her belief—her faith—she said, that compels her to do something about the data.

An atmospheric scientist, Katharine is a distinguished professor and endowed chair at Texas Tech University, the chief scientist for The Nature Conservancy—*and* an evangelical Christian. She crunches numbers, advises cities planning for the climate crisis—and she's on a mission to get people talking about climate. That means helping them connect the dots between the climate crisis and what they already care about.

As a Canadian, Katharine didn't grow up seeing a conflict between science and religion. Her father was a science educator in Toronto and also a missionary who took the family to live in Cali, Colombia, when she was nine years old. But toward the end of her undergraduate studies at the University of Toronto, she took a course in climate science that changed her life.

"I didn't realize that climate change wasn't just an environmental issue—it's a threat multiplier," she writes. After studying physics and astronomy, she saw how the climate amplified serious humanitarian issues from the refugee crisis to systemic poverty. "How could I not do everything I could to fix this huge global challenge?" she said.

With her PhD in atmospheric science, she's coauthored global reports like the US National Climate Assessments and conferred with former President Obama on climate policy. But she also watched the climate narrative evolve after moving with her husband to Lubbock, Texas, known then as the second most conservative city in the country.

Katharine reminds us that where we fall on the political spectrum in this country is the number one predictor of whether we agree that the climate is changing, humans have caused it, and the impacts are dangerous. And she wants to help frame that conversation beyond scientific journals—communicating through social media, a PBS digital series *Global Weirding*, and even an appearance on Jimmy Kimmel Live. She curates the most widely followed Twitter list of climate scientists (@scientists who do climate) as well as one for scientists who knit, craft, and paint (@scientists who create). She often brings her knitting to high-level scientific panels.

The number of Americans who think global warming is happening outnumbers those who don't by four to one (70 percent versus 15 percent), according to a recent report from the Yale Program on Climate Change Communication. Yet two in three people in the United States (67 percent) say they rarely or never discuss global warming with family or friends. So the number one thing we can do about the climate crisis, according to Katharine, is to talk about it.

Her approach to disseminating the science is radically different from technical jargon in the scientific literature: "Start from the heart," she says. Begin by talking about what matters to us—our children, the outdoors, national security, faith. Then connect the dots between those values and why we could care about a changing climate. As a Christian, she believes we have a responsibility to care for those less fortunate who bear the brunt of the impacts of global warming. I still remember the look of awe on the face of one of my students when we attended one of her talks. That same student is now a policy director at Chesapeake Climate Action Network.

When she's attacked on social media or gets harassing calls or letters—which happens often—she notes that climate change is often framed by political conservatives as a "belief system" that they feel threatens their identities. Katharine knows dialogue isn't viable with the 10 percent of people who've built their identity on dismissing climate change, but that leaves the

> "The number one thing we can do about the climate crisis is to talk about it."

other 90 percent to communicate shared values and practical solutions. "To care about climate change, you only need to be one thing," she writes, "And that's a person living on planet Earth who wants a better future."

Katharine calls this "rational hope." We need a vision of a better future, she says, of abundant energy and a stable economy, a rationale she presents in her book, *Saving Us: A Climate Scientist's Case for Hope and Healing in a Divided World*. Living in West Texas, Katharine connects with others as a mother who cares about her child, a community member where water is already scarce, a Christian, and a resident of a state that gets over 20 percent of its energy from renewables. And she's not giving up on the science or the beliefs she holds close to her heart.

VIRGINIA

TIKTOK FOR CLIMATE:
TRANSLATING CLIMATE INFORMATION ACROSS THE WORLD

SOPHIA KIANNI

CLIMATE ACTIVIST

McLean, Virginia

It could have been any of thousands of TikTok videos filmed by teens on iPhones propped on a stack of books in their bedrooms, featuring arm-popping and hip-hopping moves. But in this video, Sophia Kianni's friend danced by her bedside to electro-pop music with this text box on the screen: "How to get easy community service hours at home during quarantine." The camera quickly panned to a laptop with the Climate Cardinals website and instructions: "Sign up to volunteer. Options are provided how you can help the climate and yourself all from home :)!!"

On the first day, the TikTok video was viewed 100,000 times, and 1,000 youth signed up to help translate climate information to those who aren't English speakers across the globe. One year later, Climate Cardinals, the nonprofit started by Sophia, has engaged 8,000 volunteers in more than 40 countries translating 500,000 words about climate in more than 100 languages. Those are impressive stats for a senior in high school during a pandemic when classes were online.

"I'm Iranian American, and what I've really noticed is that there is a lack of accessibility in the climate movement to people who don't speak English," she said on *CNN*.

Indeed, the top ten most vulnerable countries to climate impacts are those where English isn't the primary language. Yet the majority of searches in Google Scholar for climate publications are in English. Among the first documents translated by Climate Cardinals was a simple climate glossary with responses to questions such as: "What is climate change? What are greenhouse gases?" The average age of volunteers is about 15, and the nonprofit partners with organizations like Translators Without Borders and proofreads documents that include technical language. Many students were eager to engage in climate action when most were quarantined at home in their rooms.

In middle school, Sophia first noticed the need for information about climate in Farsi when she traveled with her family to Iran. "I was shocked when I learned that due to climate change, temperatures in the Middle East were rising more than twice the global average," she said. But her relatives in Iran hadn't heard about the climate crisis, so she started working with her mother to translate climate news into Farsi and talk about it with them.

By the time the pandemic hit, Sophia was already active as a climate advocate, using social media and public events to draw attention to the need for action. She'd participated in a one-day hunger strike with the environmental group Extinction Rebellion in Washington, DC, protesting outside Speaker Nancy Pelosi's office and speaking at the event.

She's one of many youth climate leaders who are women, not only in this country, but across the world. Sophia is now a national strategist for Fridays for Future (the group started by Greta Thunberg), international spokesperson for Extinction Rebellion, and national partnership coordinator for Zero Hour, another youth climate organization. She named her own organization Climate Cardinals in large part because the cardinal is the state bird of Virginia.

With her many roles, she kept a diary for the publication *Refinery 29*, recording the different directions of her work. In one day's entry, she planned website content with her Climate

Cardinals team; checked in on a Google Hangouts for a PBS Fellowship, practiced a virtual lecture for Earth Day, spent some time on Pinterest, went running with her sister, and translated graphics for Climate Cardinals. With her global reach, she was selected to represent the United States as the youngest member of the United Nations Youth Advisory group on climate change.

"It's really a full-circle moment because I'm going from sharing my translations with ten of my family members to their ten million Iranian followers on Instagram," she told Forbes magazine, after securing a partnership with Radio Javan, a Persian radio station. As a member of Gen Z, she's harnessing translation to provide information about climate on social media in languages other than English, even as a student now at Stanford University studying public policy and climate.

"It's really a full-circle moment because I'm going from sharing my translations with ten of my family members to their ten million Iranian followers on Instagram."

When I watch my own teenager practicing dances on TikTok, I think about how this generation uses their devices for moves that make an impact far beyond their rooms at home.

WEST VIRGINIA

PERSISTENCE PAYS OFF:
ADVOCACY CAN FIGHT COAL AND BUILD A RENEWABLE FUTURE

MARY ANNE HITT

CLIMATE STRATEGIST AND SENIOR PROGRAM DIRECTOR

Shepherdstown, West Virginia

She will never forget their voices as she listened to people testifying about the impacts of the coal industry on their lives. "I remember being in EPA hearings in North Carolina about coal ash polluting drinking water," Mary Anne Hitt said. "Or the public hearings in River Rouge, Michigan, where people were dying of asthma attacks." In this Black community outside Detroit, it took five years of advocacy, but the utility finally closed the two coal-fired power plants operating without modern pollution controls.

"It doesn't have to be this way," she said. "We have better, cheaper ways to heat our buildings and turn on our lights. And now more than three million Americans work in the clean energy industry, three times the number that work in fossil fuels. But advocacy makes a difference."

What Mary Anne and other climate activists have done—harnessing science, national strategy, and grassroots pressure—can be measured in how the United States now gets its energy. Today, we get less than 20 percent of our electricity from coal, compared to half of our power ten years ago, and we're on track to phase out coal by 2030, she writes. By 2019, the

"In the beginning, everyone said we were crazy. No one could imagine it was even possible."

United States got more power from renewable energy than coal for the first time in 130 years.

In large part, her work with nonprofits like the Sierra Club focused on the effort to retire the 530 coal plants in the country. "In the beginning, everyone said we were crazy," she recalled. "No one could imagine it was even possible."

The 345 coal plants (and counting) that have been retired have each taken five to ten years of persistent work, including efforts to transition workers to more sustainable jobs. Her collective advocacy with the Sierra Club's Beyond Coal campaign made a difference, even in my own hometown of Asheville, North Carolina, where my students and I attended rallies, wrote letters, and lobbied to transition the local power plant off coal.

Mary Anne's relationship to the land began in her childhood home in Gatlinburg, Tennessee, the gateway to the Great Smoky Mountains National Park, where her father worked as the chief scientist and her mom was a school teacher and administrator. She grew up in a world-class biodiverse landscape with rich human culture as well. "I mean, Dolly Parton went to my high school," she said. "I grew up with a strong sense of people, place, and nature."

But she also realized that beautiful places need defenders. Around the kitchen table, she heard her father talk about how air pollution and acid rain were killing the trees in the park in the 1980s and 1990s. "I remember when he figured out that the TVA coal-fired power plants were the cause of the pollution," she said. That was at the same time she witnessed massive development in her hometown, with farms converted to shopping malls, and open spaces into gas stations.

For her it was the issue of mountaintop removal—blasting the tops off mountains to access coal—that became what she calls her "aha" moment working for Appalachian Voices, a regional environmental nonprofit. "When I flew over the mountains to see them completely destroyed and polluting people's air and water, that brought me into this work," she said. "I'd grown up revering mountains for their beauty, culture, and the sustainable economic opportunities they can provide."

With the iLoveMountains.org campaign, her staff at Appalachian Voices built a national online network to end the destructive practice of mountaintop removal—an especially egregious form of coal mining whose negative impacts affect the long-term health and ecology of Appalachian communities.

"US coal plants are built in vulnerable communities who suffer the impacts of dirty fuel," she writes in the book *All You Can Save*. She led the Beyond Coal campaign with Sierra Club for more than a decade, working with over 300 partner organizations to block new construction of more than 200 proposed coal plants and secure the retirement of more than two-thirds of the existing 530 plants. These efforts will decrease greenhouse gas emissions, save lives, and help transition the country to clean energy.

Recognizing the role of economics, Mary Anne now serves on the board of Solar Holler, a unionized solar energy company providing renewable energy and creating solar jobs in Appalachia. In 2021 she moved to a new position as the Senior Director of Climate Imperative Foundation, a nonpartisan energy and environmental policy firm, where she's working to support the biggest climate policy decisions being made in the world to reduce emissions at the speed and scale needed to tackle the climate crisis. The $369 billion investment in clean energy and climate action in the first federal climate bill is one initial part of that equation.

A primary driver for her work is her daughter Hazel, an eleventh-generation West Virginian. "I want to protect her future, but I also don't want to miss out on her life," she said. Offering insight to others involved in this work, she said: "It's not easy, and my advice is to be a little more gentle with yourself and continue to advocate."

NORTHEAST

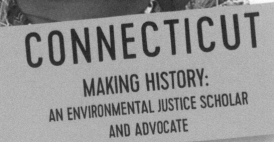

CONNECTICUT

MAKING HISTORY:
AN ENVIRONMENTAL JUSTICE SCHOLAR
AND ADVOCATE

WANJIKU
"WAWA" GATHERU

CLIMATE JUSTICE ADVOCATE

Pomfret, Connecticut

Wanjiku Gatheru, or Wawa as she's known, had firsthand experience as the only Black student in many of her environmental clubs and activities on the campus of the University of Connecticut. She'd search for faces that looked like hers in the pages of Environmental Studies texts or books like *A Sand County Almanac* and *Desert Solitaire*. It was the intersection of people and places, not just conservation, which drew her to the field.

As an undergraduate, she was part of a research lab looking at food insecurity in local communities and realized the variables measured in the surveys also applied to students she knew at the university. One of the researchers she approached about her idea wasn't supportive at all, but another mentor encouraged her to apply for funding to study food insecurity on college campuses, at that time a novel idea. Her research revealed that 25 percent of students had faced concerns about having enough food due to lack of money or other resources. The findings prompted initiatives like pop-up food pantries and garnered attention at the state and federal level about food access and racial inequality in higher education.

"But unlike many of my classmates, I did not choose my area of study on the basis of childhood curiosity. I chose it out of fear of a climate future that ignores Black lives," she wrote in a viral essay for *Vice* entitled "It's Time for Environmental Studies to Own up to Erasing Black People."

While she describes herself as an introvert, Wawa made a historic impact during her time at UConn, as students refer to the school, by being named a Rhodes Scholar, Udall Scholar, and Truman Scholar. This trifecta of academic accomplishment brought television cameras and radio interviews to campus. Her tweet says it all:

"Just when I think I've run out of tears, they just. Keep. Coming. I am a 2020 Rhodes Scholar, the 1st in UConn's history and (by the look of the archives) the first black person to receive the Rhodes, Truman, and Udall. This is unreal. Mom and Dad—I did it!"

While Wawa grew up in Connecticut, her parents had both immigrated from Kenya. She remembers her mother keeping a garden, a connection to a long line of farmers among her family. Growing up, she didn't see herself as an environmentalist, a label which seemed to describe white people who hiked or rock climbed and wanted to conserve the wilderness.

She notes that mainstream environmental scholarship doesn't center the experience of people of color but often sees it as a part of environmental justice, a sidebar to environmental studies. On social media she's making her mark on this history to grapple with a racist past in conservation as a pathway to conversation and change. The nonprofit organization she founded, Black Girl Environmentalist, is a place for "Black girls, women, and non-binary folx in the environmental movement."

"I think it's so important as a young woman and a young woman of color to be a leader in the environmental space," she said at a recent MAKERS Conference. "Because when you think of who is impacted most, who is impacted first and worst by environmental degradation and climate change, it's communities of color, it's Indigenous communities. And if we want to be specific, it's women.

"What I'm doing right now is trying to make that change, trying to organize, trying to ensure that young people know that their problems are validated in the environmental space."

As a Rhodes Scholar, she was one of thirty-two Americans selected from more than one thousand applications, although she almost didn't apply until one of her mentors said, "Wawa, why not?" She took her vision and passion to Oxford at St. John's College, where she pursued a Master of Science degree in Nature, Science, and Environmental Governance, studying obstacles to participation by people of color in the environmental arena.

Ultimately, Wawa wants to continue working toward just climate solutions and has even considered running for Congress to become the first Black Congresswoman from the 2nd district of Connecticut. Until then, she gave her perspective on this moment in history and her place in it at the MAKERS Conference: "Right now, we are at this really epic and beautiful moment where so many people from around the world are

"The advice that I'd want to give young women is something I've had to internalize myself. It's the fact that my emotions and my emotional response to injustice are not a bad thing."

saying, 'Oh my goodness, look at these youth. They're completely shifting the conversation.' The advice that I'd want to give young women is something I've had to internalize myself. It's the fact that my emotions and my emotional response to injustice are not a bad thing."

MAINE

INTERSECTIONAL MEETS INTERGENERATIONAL:
CONNECTING YOUTH AND ADULTS FOR CLIMATE JUSTICE

CASSIE CAIN

YOUTH ENGAGEMENT COORDINATOR

Portland, Maine

It's no surprise that different generations often speak a different language—a few minutes on social media reveals that age-old adage. But Cassie Cain negotiates that reality daily in her work as youth engagement coordinator with 350 Maine, the climate justice advocacy organization, and Maine Climate Action Now.

"The youth I work with sometimes have different scales of urgency when it comes to the climate," she said. "I've found young people speak with a much more personal lens: Should I have kids? What will my future look like?"

She's also noticed that youth organizers are often quick to accept the concept of intersectionality, such as recognizing that someone's race or gender affects how they experience climate change. "Young people want to integrate Black Lives Matter demands or uplift the demands of the Indigenous communities in climate work," she said, even as youth climate movements emerged within the historical legacy of civil rights and social equity.

Cassie has developed strategies to build authentic coalitions between climate organizations and youth, rather than merely putting a student's photo on a webpage or viewing them as a monolithic group. Giving leadership opportunities to students is one concrete option that can include board positions, fellowships, or paid internships, as well as compensation when asking youth to speak on panels.

> "So many of the young people go to school, play sports, and then focus on climate organizing afterwards, and they are doing the adult work of advocating for a livable future."

"So many of the young people go to school, play sports, and then focus on climate organizing afterwards," she said. "And they are doing the adult work of advocating for a livable future."

In collaboration with Maine Youth for Climate Justice, Cassie partnered with middle school, high school, and college students to organize a climate strike in Portland, an event precipitated by Fridays for the Future, Greta Thunberg's organization.

On the heels of this gathering of 2,000 people, young people, including Cassie, testified at the Portland City Council to push the municipality to declare a climate emergency. They then met monthly with the sustainability office of Portland to ensure policies aligned with the climate emergency resolutions.

"I've worked with middle school students who are on TikTok a lot, but they know more about political theory than I do," she told me. "We are young people of different ages learning from each other, from millennials to Gen Z."

Cassie grew up in the coastal town of Kennebunk and spent time with her family in a small off-the-grid cabin with no running water in the western part of the state. An environmental science class in high school piqued her interest in the climate crisis, and at Brandeis University, she spent four years in the divestment campaign, pushing the college to divest from fossil fuels. Shortly after graduation, the board voted for a partial divestment from coal, prompted by student action.

She didn't realize this experience would influence her job with 350 Maine so soon after college. As a resource person for younger organizers, Cassie worked with middle school student Anna Siegel, who wanted to resurrect a divestment bill for the state of Maine.

"A similar bill had been introduced in 2013 to divest Maine's pension fund from fossil fuels, but it had been tabled," Cassie said. "This was a stop-the-money pipeline movement." They partnered with the Sierra Club Maine, but Anna was the primary organizer for the outreach to legislators to support bill *LD 99: An Act to Require the State to Divest Itself of Assets Invested in the Fossil Fuel Industry.*

In 2021 LD 99 was signed into law by the governor, the first bill of its kind in the country. The climate impacts in the rural state include the Gulf of Maine, which is warming faster than 99 percent of global oceans, jeopardizing the lobster and shrimp fishing seasons. Human health is also at risk in Maine, given the increase in tick and mosquito-borne illness, such as Lyme disease, by an incredible 1800 percent in the last ten years. Since the elderly account for 20 percent of the state's population, intergenerational climate coalitions are essential.

But Cassie worries about the mental health of young people doing the work of adults. "I'm concerned that they are going to burn out at such a young age," she said. "We need intentional systems to support students doing this work." While continuing her 350 Maine work remotely, Cassie has begun a Master's in Social Work program at the University of Denver to provide counseling to youth climate activists. "I want to help young people continue this work and stay healthy," she said. She's seen firsthand how therapy and support systems have nurtured her own health and the health of the places she loves.

MASSACHUSETTS

BUILDING A YOUTH CLIMATE MOVEMENT:
THE RIGHT TO GOOD JOBS AND A LIVABLE FUTURE

VARSHINI PRAKASH

EXECUTIVE DIRECTOR

Boston, Massachusetts

These students weren't in Washington, DC, for a field trip. Instead, they came to fight for their future. Wearing bright yellow and black T-shirts, they sat shoulder-to-shoulder on the blue carpeted floor of Speaker Nancy Pelosi's Congressional office. Some held yellow banners with statements like "Step Up or Step Down." One young man in his wheelchair advocated not for the planet, which will persist, but for his generation, which might not.

Two hundred young people, from seventeen to thirty-five years old, gathered in Speaker Pelosi's office as members of the Sunrise Movement, the largest youth movement with the aim to stop climate change and create millions of good-paying jobs. This was one way to get the attention of the adults in the room. They were joined by then Representative-elect Alexandria Ocasio-Cortez in this made-for-social-media moment to challenge both Democrats and Republicans to transform our economy and society and avert the climate crisis.

As the cofounder of Sunrise in 2017, Varshini Prakash knows the power of combining big viral moments with routine, consistent action in local communities. Teenagers volunteering with Sunrise made 850,000 of the 1.3 million phone calls to help Representative Jamal Bailey, a progressive candidate, secure his seat in the New York State Senate. Sunrise, with hubs in regional and local centers, trains young people to initiate and leverage action to release the hold of the fossil fuel economy and corporate dollars on our country. They are leaders in a fight to change this unsustainable status quo.

> "The most effective way to get people involved is to ask them to join. The second is to ask, What's at stake? What is their skin in the game? What do you love that you stand to lose in the climate crisis?"

"The most effective way to get people involved is to ask them to join," Varshini told *NPR*. "The second is to ask, What's at stake? What is their skin in the game? What do you love that you stand to lose in the climate crisis?"

As the daughter of South Indian immigrants in Boston, Varshini didn't see herself as someone with power. "Growing up as a Brown, skinny, short girl, I felt that the whole culture of politics and elections and the government reveled in my exclusion," she told journalist Ezra Klein. She remembers feeling powerless as an eleven-year-old watching news of the tsunami in the Indian Ocean, followed later by Hurricane Katrina, and then disastrous flooding in her father's hometown in India.

It was when she was in college that she fell in love with the power of social movements. When she emceed a demonstration against fossil fuel infrastructure in Massachusetts, she began to see herself as a catalyst for other students fighting for their generation's livable future. Holding the megaphone and looking out at the crowds, she felt connected to a larger cause. This led her to participate in direct action, including protests against the Keystone XL pipeline.

"When I was in high school, I had low self-esteem," she said. "I didn't believe I was powerful or smart or worth investing in." But organizing changed her—and it changed the political landscape for climate action and youth in this country.

In 2016, with a network of young activists, she helped create a youth climate movement after spending one year studying effective social movements across the globe, especially the US civil rights movement, built with the participation of youth. As they researched the past, the young leaders recognized the climate crisis was becoming a full-blown emergency with no political home for the youth who cared about it.

At the time Sunrise was founded, climate change wasn't in the headlines in this country, despite disasters such as hurricanes, wildfires, and drought as daily realities. Now through storytelling, singing, digital organizing, in-person trainings, and action, Sunrise has elevated policies like the Green New Deal into the everyday language of elected officials.

I've watched my own students participate in Sunrise events, including organizing a sit-in at the mayor's office in Asheville, North Carolina, that pushed the city council to declare a climate emergency. As Varshini often says, "We do civil disobedience so we can bring the private suffering that is happening across the country to the public into everyday dialogue."

For the Sunrise movement, the power of engagement comes from three sources: people power, political power, and finally, political alignment, bringing a shared agenda for society with the Green New Deal. As such, the Green New Deal provides a ten-year mobilization plan to tackle climate change and inequality in this country and goes beyond the provisions of the federal climate legislation passed in 2022.

"We are ready to be the adults in the room," she often says. "We are ready to envision reality in a different way."

NEW HAMPSHIRE
WHAT WE KNOW FROM SNOW:
WHY SKIERS AND SNOWBOARDERS
SHOULD CARE ABOUT CLIMATE

ELIZABETH BURAKOWSKI

CLIMATE SCIENTIST

Durham, New Hampshire

In her early thirties, Liz Burakowski scaled a 120-foot tower to collect data on snow reflectance, or "albedo," the proportion of light reflected rather than absorbed by the frozen accumulation. What her research revealed was the crucial role of snow to the environment: snow helps maintain and build water supply, provides a habitat for wildlife, and keeps the ground from freezing, which affects microbial systems that promote decomposition. When regions experience decreased snowfall due to global warming, the change throws everything out of balance, often with disastrous consequences, like reduced snowpack and years-long drought.

But it was her studies on the economic impacts of global warming on the ski industry that resulted in calls from newspapers and television.

"Once my research focused on jobs, people started paying attention," said Dr. Elizabeth Burakowski, known to friends as Liz.

As a research professor at the University of New Hampshire's Institute for the Study of Earth, Ocean, and Space, she uses climate models, snow measurements, and satellites to understand how changing winters affect the northeastern United States. Her study on the impact of warming temperatures on winter tourism, a $12.2 billion industry in this country, showed how climate affects job loss.

It turns out low-snow years can result in a $1 billion decrease in economic activities and the loss of 17,400 jobs. These findings—especially relevant in the small mountain towns of northern New Hampshire reliant on skiing and snowboarding—revealed the significance of 65 degree rainy winter days deterring skiers and riders. "If you have back-to-back warm and rainy winters, it can also be hard to keep people interested in the sport," Liz said.

Her obsession with snow began when she was seven years old, and her family moved from Wisconsin to New Hampshire. As a tow-headed child bundled in a snowsuit, she learned to ski with her siblings at the small, affordable local resorts, where lift tickets were $10 and used ski equipment cost $20.

"I remember feeling the thrill of going down a hill and asking, 'When can we do that again?'" she said.

In the 1980s, she began skiing when the levels of carbon dioxide, a greenhouse gas, were 350 parts per million (ppm). This number reflects how many parts of carbon dioxide are in a million parts of air. Scientists agree 350 ppm is the safe level needed for our life on this planet.

In 1995 Liz took up snowboarding when carbon dioxide levels were 360 ppm, and by the time she started graduate school ten years later, the levels had climbed to 380 ppm.

"My students just took their final exam and CO_2 levels were 405 ppm and expected to reach 500 ppm in our lifetime," she explained in a TEDx Talk. She's been skiing for thirty years—a time period that has seen winter temperatures in the northeastern United States rise 5 to 6 degrees.

Winter is the fastest warming season across much of the snow-covered United States. Given this reality, 600 of the small ski resorts in the country have closed. Fewer than 500 remain, which often means they have to make snow. In the northern counties of New Hampshire, 45 percent of visitor spending from tourism comes from the ski industry. So a bad year can devastate local economies.

When Liz was a geology major at Wellesley College, she would watch the weather in hopes of skiing on a weekend. But often, there were 60–degree days in the winter. "It felt wrong," she said. "That's really what got me started in the study of climate change."

While she studies the impact of snow on ecosystems and economies, she's also looking at examples of businesses and municipalities cutting emissions such as the University of New Hampshire, which operates on 100 percent renewable energy.

Now her son is six years old and looks forward to skiing as she did as a child. Her two-year-old daughter did an intentional face-plant into snowy drifts to taste the snow before she could even walk. "Snow is important for ecosystems, but winter is also something I want to preserve for myself and my four-month-old son," she wrote in *The Guardian* soon after she became a mother.

> "Snow is important for ecosystems, but winter is also something I want to preserve for myself and my four-month-old son."

A recent study she coauthored showed that fewer than half of New Hampshire survey respondents were aware that winters are warming across the state. In a warming climate, regular cold feels colder by comparison, which reminds Liz of the term "solastalgia," the sense of loss when your environment is changing. She describes the experience—"like something is off"—when she's wearing shorts and a T-shirt in February in New Hampshire.

As a scientist, Liz recognizes the impacts of global warming on the ski industry are symptomatic of a bigger climate crisis that disproportionately affects those without resources to adapt. But for small towns that depend on tourism—and those who value winter recreation—these shared values are a lifeline to culture and place.

NEW JERSEY

SPEAKING OF FAITH:
WHAT RAMADAN REVEALS ABOUT THE CLIMATE AND OUR LIVES

SAARAH YASMIN LATIF

ENVIRONMENTAL EDUCATOR AND COMMUNITY ORGANIZER

Newark, New Jersey

When most people consider the holy month of Ramadan, the thirty days of fasting and reflection for Muslims, they may not picture a millennial in a hijab connecting the Qur'an with environmental justice through Instagram hashtags like #greenramadan and #ecomuslim. But Saarah Yasmin Latif is on a mission to help people of all religious traditions connect their faith with individual and collective acts to sustain the earth.

During Ramadan Muslims fast from dawn to sunset as part of their religious obligation. Fasting is meant to draw Muslims nearer to God with a focus on discipline, gratitude, and compassion for those less fortunate. Ramadan is also known as the month in which the Qur'an was revealed to Prophet Muhammad (peace be upon him, which Muslims say as a sign of respect).

When Saarah created the Green Ramadan Challenge, she included three parts to each daily post: words from the Qur'an, an environmental challenge, and a reflection. For the first day of Ramadan, she chose these words: "Did they not look at the earth—how much we have produced therein from every noble kind?" (Qur'an 26:7). The challenge was to spend five to ten minutes outdoors in nature. Leave your phone inside. Listen to the rhythm of the earth. Pay attention to the sounds of the trees, animals, and wind. And the reflection asked: Allah created us as the caretakers of this Earth. What are you personally doing to fulfill this trust?

"The Instagram posts formed a tight-knit community," Saarah told me. "In the comments, people shared how they were taking action. In Ramadan I feel a sense of tranquility and strive to be the best person I can be and to connect that to the environment."

From her home in Newark, Saarah focused the first virtual Ramadan challenge on building community through small, consistent acts in 2020. During the pandemic, Muslims were breaking the fast (*iftaar*) without the communal support of their Masjid, or mosque. Saarah's eco-challenge, as she also called it, revealed her belief that we are placed on Earth to be stewards and to care for each other and the environment.

In the suburbs of New Jersey, Saarah was the child of immigrants from the Caribbean country of Guyana. Her parents encouraged her to play outdoors until the sun set in their neighborhood in South Orange. "In my teenage years, I realized how drawn I was to nature," she said. "As a millennial, I feel like I am a part of the last generation who grew up without screen time as a constant."

While she was raised in the Muslim faith, she didn't connect her religion with the environment in her youth. In college, however, she enrolled in a study abroad course in New Brunswick, Canada, that nurtured her interest in local environments and communities, leading her to pursue a master's degree in sustainability and leadership.

> "Yet the closer I grew to creation and nature, the greater my love for God became."

"I always saw them as two elements, God and creation," she said. "Yet the closer I grew to creation and nature, the greater my love for God became."

Those intersections led to a five-month internship focused on urban farming with BIPOC communities. She has led more than twenty sustainability workshops, both in-person and virtual, within her local Masjid in Newark and surrounding communities, often integrating simple actions like bringing reusable plates and silverware to break the fast in Ramadan, rather than using single-use disposable plastics.

With her emphasis on interfaith dialogue, Saarah continued to ask the question: Why aren't Muslims on the forefront of the environmental movement, especially since environmentalism is embedded in this religion established more than 1,400 years ago? For answers, she turned to the sayings and life of the Prophet Muhammad (peace be upon him), which are called "Hadith" in the Muslim faith.

As a fellow with a program called GreenFaith, Saarah collaborated with participant Kori Majeed on an anthology of writings reflecting environmental justice. More than a decade ago, I'd also participated in a GreenFaith fellowship and seen the power of interfaith dialogue for the climate.

"We wanted a document that could be used in interfaith gatherings," she said.

The result is a free and accessible e-book, *Forty Green Hadith: Sayings of the Prophet Muhammad (peace and blessing be upon him) on environmental justice and sustainability*. This book allowed her to grow a network, collaborating with other Muslim-led environmental groups—and sharing content together through social media. When she isn't writing or joining a climate justice protest, she's integrating nature into the learning center where she teaches, another way to live out her faith.

"I find myself not just an admirer of the hues of leaves, the stillness and resilience of trees, and persistence of roots to blossom, but as an avid advocate for the protection of the world we live in," she writes.

NEW YORK

OCEAN JUSTICE:
A MARINE BIOLOGIST BUILDS
A BIGGER TEAM FOR THE CLIMATE

AYANA ELIZABETH JOHNSON

MARINE BIOLOGIST

Brooklyn, New York

"I'm curious if there was a moment, Billie, when you were like, 'Oh shit, climate change?'" With her signature gold hoop earrings and black-rimmed glasses, Dr. Ayana Elizabeth Johnson bantered about the climate with Grammy-winning singer and songwriter Billie Eilish on Nike's sustainability series "Talking Trash," which used conversation and compelling graphics to explore the climate crisis for its ten million young viewers.

During the dialogue Ayana explained that addressing climate change will involve a transformation of everything, from agriculture to transportation. "It helps me to choose one area of focus, and since I'm a marine biologist, I'm focusing on the intersection of climate and oceans."

For Ayana, ocean justice is where ocean conservation and social justice intersect, even if many people aren't aware of the role of oceans in the climate system. She calls the omission of ocean conservation in climate solutions the "big blue gap," especially given that oceans are becoming hotter and more acidic, fueling more intense hurricanes and storms. Yet oceans also hold large-scale promise for climate solutions. When Ayana first read the Green New Deal, the resolution aimed at transforming our fossil-fuel energy system and creating jobs, she noted a glaring omission: there was only a passing reference to the world's oceans.

"Our nation has more than 95,000 miles of shoreline, home to 40 percent of Americans who live in coastal cities," she wrote in the *Washington Post*. "Our blue economy, including fishing, ocean farming, shipping, tourism, and recreation, supports more than 325 million jobs."

> "It helps me to choose one area of focus, and since I'm a marine biologist, I'm focusing on the intersection of climate and oceans."

Presidential candidate Elizabeth Warren had invited her to craft a Blue New Deal with her campaign, a recognition of her work that includes the coastal cities think tank Urban Ocean Lab as well as a decade of community-based conservation in the Caribbean. She sees the Blue New Deal as a blueprint for restoring coastal ecosystems, sequestering carbon, investing in renewable offshore energy, and supporting green jobs.

As a five-year-old girl, Ayana discovered the magical world of the ocean when she traveled with her parents from Brooklyn to Florida, where she first learned to swim. During a ride on a glass-bottomed boat, she joined the other children in throwing cheese popcorn to the fish. Because of an allergy to dairy, her body broke out in hives, and, after rinsing her off, her mom took her to the cabin of the boat. There, she had a wide-angle view of coral reefs and colorful fish and fell in love with the world under water.

She feels lucky she learned how to swim at an early age on this trip to the coast. "So few Black kids learn to swim," she wrote in a Father's Day Instagram post, "70 percent don't know how." The accompanying photo showed a young Ayana in a royal-blue one-piece bathing suit preparing to dive to her father, standing with open arms in a hotel pool.

"Our country's generations of racism and exclusion—of barring Black people from beaches and pools, is deadly. And certainly contributed to the lack of diversity in marine biology," she wrote.

As a biracial women with a white mother and Black father, she later learned from her parents that when her dad, an architect from Jamaica, joined her in the pool, the other white hotel guests refused to get into the water. This was Florida in 1985, less than four decades ago.

Yet it was that trip that fueled her passion for marine biology. Years later she earned a PhD from the University of California Scripps Institution of Oceanography, where she went on to design a fish trap to reduce bycatch in Curaçao and Bonaire, followed by studying community-based sustainable management of coral reefs in the Caribbean.

During her studies, Ayana interviewed hundreds of fishermen and divers, recognizing that ocean conservation is as much about people as coral reefs. Her public work at the intersection of gender, race, and ocean justice continues as she uses her voice in news outlets, such as a *Washington Post* op-ed after the death of George Floyd: "I'm a Black climate expert. Racism derails our efforts to save the planet."

She's building a bigger and better team to dismantle racism and sexism in the climate conversation, with ocean justice at the forefront. In her coedited anthology, *All We Can Save* with Dr. Katharine Wilkinson, she calls for a "feminist climate renaissance" that evokes creativity and rebirth.

Her conversation with singer Billie Eilish and a recent TED Talk about how to find joy in climate action are just two examples of using creative strategies to fuel collaborative change. "I'm really excited about the youth climate movement being furiously, passionately, and brilliantly led by teenage girls." She acknowledged to the teen idol that no one knows if their actions might actually tip the scales. "On my gravestone, I will earn the words: She tried. And that's kind of all we can do."

PENNSYLVANIA

IT'S FOR OUR CHILDREN:
FIGHTING FOR HEALTH AND CLEAN AIR

MOLLIE MICHEL

DEPUTY DIRECTOR

Philadelphia, Pennsylvania

"When the weather is hot, sometimes we can't play outside, and it can be hard to breathe for my friends who have asthma," said fourth grader Zada on Facebook Live with her state senator. "What should we be working on to fix climate change in Pennsylvania?"

Her mom, Mollie Michel, sat next to her and nodded gently as Zada posed questions about the impacts of climate from a kid who liked to go to circus class and play outside. After reading from her notes, Zada looked into the computer screen with what felt like a mix of relief and pride, and I watched her mom exhale with a smile.

"When my kids are advocating for themselves, there's that proud-mom feeling, but it's intermingled with the urgency of the climate crisis and the disbelief that so much is at stake— that they have to do this," Mollie told me. She was speaking about her role as a mother and a former program manager for Moms Clean Air Force, with more than 1.2 million members nationwide.

"When I was their age, I couldn't have told you the names of my elected officials," she said. "But they are comfortable explaining to their legislators why issues matter to them. I wish it wasn't so urgent though, the need for them to be heard."

Both her daughters, Zada, ten, and Caia, twelve, have joined Mollie in her advocacy through her job supporting the fifteen regional chapters of Moms Clean Air Force, including strategic work in Pennsylvania, where the governor committed to reducing greenhouse gas emissions by 80 percent by 2050. But she's aware a pledge is only the beginning. Pennsylvania, a fossil-fuel rich state, is the third largest producer of greenhouse gas emissions in the country.

> "When my kids are advocating for themselves, there's that proud-mom feeling, but it's intermingled with the urgency of the climate crisis and the disbelief that so much is at stake—that they have to do this."

During the Facebook Live, Democratic State Senator Katie Muth, who represents southeastern Pennsylvania, answered the question from Mollie's daughter by acknowledging a lack of political will within the state legislature to protect people from the impacts of the oil and gas industry.

When Mollie moved to south Philly from Brooklyn seven years ago, her kids were young, and she worked as a birth doula. She started getting involved in the local schools due to indoor air quality concerns in what she calls "their majority-minority neighborhood." In Pennsylvania, asthma rates are two times the national average, with disproportionate rates for Black and Brown children.

"As a white upper-middle class woman, I had to learn when to mute my own voice," she said. "And at the same time, I was learning how to lift up voices of those impacted by cement factories, fracking, coal-fired power plants, and poor air quality."

Several years ago she brought her children to a rally in Washington, DC, protesting the confirmation of Scott Pruitt as EPA administrator. Although she'd never been inside the Capitol Building before, she remembers thinking: "I could do this every day." After her return home, she came on board as a field organizer before moving into federal work.

Working as a birth doula was the role that taught her the most about advocacy. It was parenthood—and motherhood especially—that showed her how to connect those advocacy skills to the climate. Now Mollie and her husband, a social worker, think carefully about how to frame climate change for their daughters, who care about a range of issues, from LGBTQ+ rights to single-use plastics.

"I try to focus on things they can do and not overwhelm them so much that they become overly anxious and can't do anything," she told me. "I also talk a lot about pollution, and that sometimes evolves into talking about global warming. So when we get up at 6 a.m. so my oldest daughter speaks at a rally in Harrisburg, that is a part of the solution."

To exert that pressure for solutions, Moms Clean Air Force has worked on issues such as methane regulations for the past decade. "Here in Pennsylvania, methane is eighty-four times more potent than carbon dioxide," she said. "Methane pollution contributes to 25 percent of climate change."

With oil and gas operations in Pennsylvania emitting more than one million tons of methane each year and the current state laws leaving 50 percent of methane emissions unregulated, Mollie and others are pressuring legislators to close the loop in these protections. The state now provides standards for new sites, but none for existing infrastructure. Leveraging her experience with Moms Clean Air Force, she moved into a new position at Environmental Protection Network, a nonprofit that builds the capacity of environmental agencies and communities to confront health and environmental crises.

"We have to transition off fossil fuels, and until that happens, we have to protect our children from the oil and gas industry," she said. "I'd love for the future to seem bright for my daughters. I don't think that's too much to ask."

RHODE ISLAND

CLIMATE CHANGE IS NEITHER REPUBLICAN NOR DEMOCRAT:

IDENTITIES AND OUR CLIMATE FUTURE

EMILY DIAMOND

PROFESSOR OF COMMUNICATION STUDIES

South Kingston, Rhode Island

Emily Diamond would like you to meet three people who might influence how you think about climate change in this country. The first person is Don, a soybean farmer from Illinois, father of three, and member of his county's Republican party. The second is Debra, a retired nurse in North Carolina who's politically conservative and a leader in her local Audubon Club. Lastly, there's Cheryl, a mother of three in a Cleveland suburb and active in her church and PTA.

What could these three people, with a mashup of beliefs and backgrounds, have in common? All have become active advocates for climate change and clean energy policies. As an assistant professor at the University of Rhode Island, Dr. Emily Diamond spearheads the research to understand why.

"The multidimensional nature of our identities is part of the solution to dissolving the partisan divides around the climate," she said.

Emily's childhood in two regions of the United States shaped her interest in framing climate issues beyond the more predictable lens of political beliefs. She lived in Seattle, Washington, until she was twelve, surrounded by those who spent time outdoors and supported environmental policies. In middle school she moved to a suburb outside Atlanta, Georgia, known for being more politically conservative.

"The multidimensional nature of our identities is part of the solution to dissolving the partisan divides around the climate."

"I realized that the people I met in Georgia were just as appreciative of the outdoors, but they thought about politics differently," she told me. "In both Seattle and Atlanta, people cared about the quality of the air and water, but many people in Georgia weren't voting to support those policies."

However, in both regions she also noticed a common denominator: "People were generally dismissive and even judgmental of those from the other place," she said. "It felt like they were talking past each other."

Even at a young age, she could see how shared values—such as an appreciation for the outdoors or concern for clean air—became muddled by political views, causing those on both sides to lose sight of common environmental concerns. These early experiences shaped her work in communications after college, where she helped Fortune 500 companies understand how they could maximize both financial gains and sustainability. During her PhD studies at Duke University, she dove deeper into research on personal identities—and how they shape our perceptions of climate.

"For the most part, people's political identities are not the strongest, most salient part of who they are," she said. "But climate is strongly associated with political affiliation. So if we can think about other identities such as hunters, parents, or Rhode Islanders, we can find ways to consider how that identity might be threatened by climate change."

As we talked, I thought about my own identity as a mother and how climate change threatens the future for my daughters. But I'm also a person of faith, sharing values around the climate crisis with members of my Episcopal church. According to Emily, identities are among the most powerful drivers of human behaviors and attitudes, but they are nuanced.

"My research has shown that attitudes and behaviors toward climate policy can shift once the issue is related to another important part of a person's identity," she said.

For the past several years, Emily has taught at the University of Rhode Island in both the Department of Communication Studies and the Department of Marine Affairs. With 400 miles of coastline, Rhode Island was the first state to surpass two degrees Celsius of warming in the lower forty-eight states.

"The coast is so vital to who we are," she said. "But we know the beaches could be gone in ten years, and fish stocks are declining in a place with such a huge fishing industry." The practical application of her research involves a concept called framing, something advertisers do all the time to market products to specific audiences, such as senior citizens or millennials.

"We need to develop communication strategies to match the frame to a salient identity," she said. She talked about this strategy as "micro-targeting," including customizing communications based on sense of place, such as rural Americans or coastal Americans. "Another key concept is getting more people to talk about climate change with others who share identities that aren't political, such as religious communities or parenting groups," she said. "We all have so many aspects to who we are that are impacted by the climate crisis."

For Emily, her identity as a professor is just one part of her life. When we spoke over Zoom, she sat in a coffee shop while her husband walked outside with their newborn, Lily. Our identities reflect the essence of who we are and the future we want to see.

VERMONT

TELL ME A STORY:
BIKING AROUND THE WORLD
FOR WATER AND THE CLIMATE

DEVI LOCKWOOD

REPORTER AND EDITOR

Upper Valley, Vermont

A blue couch in her family's home in Vermont in a pandemic seemed like an unlikely location for writing a book about traveling around the world. But after her day job as an editor, that's where Devi Lockwood parked herself each night until she fell asleep at her laptop. Curled up on the L-shaped sofa, she listened to hundreds of hours of interviews—from New Zealand to New Orleans—conducted during her five-year journey to bike across the globe, seeking stories of the real-life impacts of climate change.

Through listening to audio recordings and reading her journals, she revisited encounters with the people she met facing floods, droughts, hurricanes, and other disasters wrought by the climate crisis.

"I believe water and climate are the defining issues of my generation," she told me.

As she biked she wore a cardboard sign that hung around her neck by a polka-dot ribbon, or sometimes even a shoestring. In black marker she'd written the words: "Tell me a story about water" on one side and "Tell me a story about climate change" on the other.

In 2014 when she graduated from college, Devi received a $22,000 grant to kick off what she thought would be a one-year journey that morphed into half of a decade of her twenties. She first joined thousands of young people demanding action for the climate at the People's Climate March in New York City. Wearing her sign, she heard stories about the impacts of a changing climate on the maple syrup industry in Vermont and the devastation of Hurricane Sandy in New Jersey.

"I believe water and climate are the defining issues of my generation."

"I thought I would create a sound map where people could click on a location to hear each person from that first year," she said, "But once I started collecting stories, I just kept going."

Five years later she'd crowdsourced funds to document climate stories from places as diverse as Canada, Fiji, Thailand, Laos, the UK, Australia, New Zealand, and Peru, visiting twenty countries on six continents, stories that became the framework for a book.

As a child in New England, Devi saw her mom as a role model in adventure, a mountaineer sponsored by National Geographic and whom she describes as "underprotective." When her mother left to climb Everest or K2 in the Himalayas, Devi's grandparents took care of her, not realizing she would someday inherit her mother's legacy.

While in college she first began collecting stories of climate and water as she cycled 800 miles along the Mississippi River for her senior thesis in Folklore and Mythology, recording fifty hours of interviews from the road.

"I didn't really know what I was doing," she told me. "But the more stories I heard, the more I saw the role of water and climate together," she said. "And I became a better listener in the process."

Asking people about water was often an easier way to get them talking—and it all led back to climate, and ultimately to the publication of her book organized by continents: *1,001*

VERMONT

Voices on Climate Change: Everyday Stories of Flood, Fire, Drought, and Displacement from around the World. Her target of 1,001 tales was inspired by Shahrazad, a storytelling hero in Arabian folklore who tells this many stories to a murderous king in order to save her life.

Often, Devi would discover stories along the way to her destination points, such as when she stopped at a fruit stand in New South Wales, where a farmer named Terry described how unpredictable floods increased his risks with every planting. In the coral atoll nation of Tuvalu, where she spent a month, many residents had decided to move to Fiji due to the rising seas and well water that had become too salty for drinking.

In 2017 she returned to the People's Climate March, this time in Washington, DC, where she once again carried the sign: "Tell me a story about climate change." She'd sought shade in front of the White House on a brutally hot day when a man approached her to share how he'd changed his perspective on the climate.

"I come from a conservative background," he told her, "where climate change is like a dirty word. But through my travels, I've seen how innocent people are suffering from disasters, and Jesus cares for those who are suffering." He'd created a new narrative, much as her stories do for those who may not write climate policy, but whose lives are upended by climate reality.

After traveling the world, Devi completed a year-long editorial fellowship at the *New York Times* Opinion section, later followed by work as an editor at the *Philadelphia Inquirer.* Through listening, writing, and editing, she is crafting conversations about climate and water and elevating stories that matter.

MIDWEST

MIDWEST: The States and the Stories

Illinois	Michigan	North Dakota
Indiana	Minnesota	Ohio
Iowa	Missouri	South Dakota
Kansas	Nebraska	Wisconsin

ILLINOIS

THE GREEN NEW DEAL AND BEYOND:
ALL FISCAL POLICY IS CLIMATE POLICY

RHIANA GUNN-WRIGHT

ARCHITECT OF THE GREEN NEW DEAL & POLICY DIRECTOR

Chicago, Illinois & Washington, DC

Like many twenty-somethings, Rhiana Gunn-Wright needed to find a job. Armed with her PhD, her resume was beyond rock solid. She'd studied at Oxford as a Rhodes Scholar, served as an intern with Michelle Obama, and worked as a policy analyst for the Detroit Health Department as well as for a gubernatorial campaign. Little did she envision that her next step would actually involve writing a document for a climate mobilization to restructure the entire US economy and reduce emissions. But in 2018 that's exactly what happened: the plan she wrote would soon become known as the Green New Deal.

"The truth is I was scared, and I needed a job," she later recalls in *All We Can Save*. She had reached out to Alexandria Ocasio-Cortez's campaign manager in search of a short-term job and was offered a position at New Consensus, a think tank in Chicago charged with developing the Green New Deal with the intersectional goals of zero waste, zero poverty, and zero racial wealth gap.

In an interview with Bloomberg, Rhiana described the challenge posed to her by New Consensus: "You can write this thing in four months, right?"

"No, but maybe if you give me two research assistants!" she responded. Sharing that memory, she laughed, marveling at the audacity of her own sense of possibility. But she also felt a real sense of urgency despite feeling daunted by her relative new role in the climate arena.

"I feel like often in climate policy—that if you don't have a background in those things, you can't talk about it," she said. "But the fact is that most people experience climate change in economic effects and health effects. They experience it as Hurricane Katrina, they experience it as wildfires in Paradise, they experience it as a heat wave when they don't have money to afford air conditioning."

As a native of the South Side of Chicago, in a neighborhood called Englewood, Rhiana grew up experiencing how race and inequality intersect with health. When she was a child, she had asthma, the result of air pollution. Raised by her mother and grandmother, she remembers how hard it was for her mom to take off work to stay with Rhiana if she had to miss school due to illness. As Rhiana became older, she realized that she was not alone—many of her friends in the neighborhood had asthma too.

"It wasn't natural," she said. "It wasn't that I just had lungs that didn't function the way they were supposed to. I was poisoned."

We know the impacts of climate change are felt disproportionately by low-income communities of color, and that's by design through the systemic location of coal-fired power plants and hazardous waste facilities in these neighborhoods. As a policy analyst, Rhiana brings her lived experience to the work of transforming systems of power in this country so that these communities aren't threatened because of where they live.

As the architect of the Green New Deal, she saw House Resolution 109 introduced by Representative Alexandria Ocasio-Cortez and Senator Ed Markey as a ten-year plan with five goals: achieve net zero greenhouse gas emissions through a just transition, create millions of high-wage jobs, invest in infrastructure and industry, secure clean air and water, and promote justice and equity by repairing historic oppression of frontline and vulnerable communities.

Organizations collaborating in a Green New Deal Network aim to create an economic mobilization on a scale which this country hadn't seen since World War II. While the initiative was controversial at the outset, the Biden Administration took many of the ideas in the Green New Deal into its own climate proposals.

With her expansive vision, Rhiana is someone who sees the connections between police brutality, community health, and the climate. As the director of climate policy for the Roosevelt Institute, she became the lead author on another proposal, this time for climate-forward stimulus policies in response to the pandemic.

"We argue that all fiscal policy is climate policy."

"We argue that all fiscal policy is climate policy," she writes in *A Green Recovery*, noting that the CARES Act supported the fossil fuel industry, which claimed more than $5.8 billion in aid. In contrast, the European Union allocated 25 percent of economic recovery to initiatives like renewable energy. She's voiced her opinion on Twitter about the challenge of holding "two truths" of the federal climate legislation: the need for investments in clean energy now and the harm to frontline communities from the continued support of fossil fuel industries.

"It's also about changing power relationships," Rhiana said. "The ability to use fossil fuels without restrictions requires people who are seen as disposable."

A generation ready to rewrite social and economic systems is not ready to write people off. She reminds us that 78 percent of Americans live paycheck-to-paycheck, an inequity exacerbated by the climate crisis. Through policy, she wants to create right systems for the good of all—from the South Side of Chicago to the halls of Congress.

INDIANA

DO YOU KNOW ANYONE
WITH ANXIETY?

AN ECOLOGIST HEALS RELATIONSHIPS WITH THE LAND

LOU WEBER

ECOLOGIST

Fort Wayne, Indiana

A seasoned professor, Dr. Lou Weber greeted her ecology class on the first day and quickly realized she had a problem. Every single student continued to stare at their phone, perhaps texting friends or scrolling through TikTok, even after she welcomed them. At the University of Saint Francis, all pre-med students were required to take this 8 a.m. course, another strike against her.

"So I took my notes, threw them into the wastebasket, and started asking questions," she said. "Does anyone know someone who is anxious?" A few students looked up from their devices.

"Does anyone have a friend or family member who is depressed?" Heads nodded. So she pivoted to asking questions about the land.

"Does anyone think they'll ever own land?" Several raised their hands.

"Would anyone be interested in learning wilderness medical skills?" At this point, students took out their notebooks, poised to learn.

"I had them at this point," Lou said. "These questions were all connected. I knew in that moment I had to change how I taught ecology to focus more on healing relationships between students and the land."

Once she had their attention, she began discussing how nature can change our physiology to reduce stress. The course she teaches still focuses on field work and ecological concepts such as biodiversity and climate change, but Lou's priority is their relationship to place, even in rural Indiana. She's incorporated activities as simple as observing the landscape while walking between classes, building a shelter from debris, starting a fire, and reducing screen time.

"I knew in that moment I had to change how I taught ecology to focus more on healing relationships between students and the land."

"The fire-starting activity has become almost legendary among the pre-med students," she laughed. "I grade them on their ability to start a fire, which is actually hard in the Midwest on a humid day. They've even torn out pages of the textbook I wrote to try and get it started. But those who fail come back years later to tell me they've learned how."

After the course, the students often share that they continue to go to nature for solace and to restore themselves.

"If we want to look at why people feel depressed and anxious, some of the reasons are staring us in the face," she said. "We have an Earth system degrading before our eyes. We've lost 20 percent of our species since 1900, and we face climate destabilization."

Both biodiversity loss and climate change share one element in common: carbon. This element forming the basis of life is part of the makeup of carbon dioxide and methane, two greenhouse gases. In Fort Wayne, Lou engages students and community members in restoring habitats for tallgrass prairie, grasses that grow ten feet high and ten feet underground, and can sequester more carbon than a lawn or even a forest. She's seen how individuals share their own stories when working outdoors to preserve biodiversity at a larger scale.

"We're restoring prairie ecosystems on campus and with community groups, like the Southwest Conservation Club and the Miami Tribe," she said. "When people work outside together, they spill their guts. It becomes restorative on multiple levels."

Her love of the land began as a child during family reunions on Beer Lake—near the town of Silver Lake, Indiana—with 200 extended family members in a tradition dating back to 1927.

"I thought I was going to heaven for five days, being in nature with our huge family and celebrating rituals together," she said. "It was easy to be drawn to conservation."

But she was also connected to her Catholic faith and spent two years in the late 1980s discerning a call to become a nun with the Sisters of the Holy Cross in Notre Dame, followed by the Sisters of Loretto in Colorado.

"During that time, I was volunteering at a homeless shelter in Colorado," she told me. "And I couldn't decide whether to continue this life of service or become an ecologist."

She described her dilemma to some of the residents at the shelter, who'd just watched a television show about climate change.

"Girl, if you are being called to be an ecologist, go do it," they told her. "Anyone can run this shelter, but we need you to save this world."

Following that counsel, she enrolled in a PhD program, followed by a thirty-year career in teaching, including a decade when she chaired my department at Warren Wilson College in North Carolina. And she's not done learning yet. She's pursuing a master's in clinical mental health counseling with a specialty in eco-therapy, and plans to complete her degree before she retires from teaching in a few years. In the meantime, she's published "An ecologist's guide to nature activities for healing" in the journal *Ecopsychology*, inspired by the Indigenous concept of reciprocity: healing ourselves by healing the land.

IOWA

NATURE AS NURTURE:
A PRESCRIPTION FOR THE HEALTH OF ALL

SUZANNE BARTLETT HACKENMILLER

PHYSICIAN

Cedar Falls, Iowa

During an especially busy workday as an OB-GYN, Dr. Suzanne Bartlett Hackenmiller found herself looking at the trees outside her exam room, yearning for an alternative way to heal.

"Don't you wish we could take this outside?" she would sometimes ask her patients. "They would look at me as if I was crazy—remember, I'm a gynecologist!" she told me.

Still, she held onto that question. More than a decade ago, she met with a nonverbal adult patient with autism whose parents had brought her to the clinic. "When I walked into the room, she was agitated, and the parents and a caregiver from the group home were trying to calm her," Suzanne recalled. "I asked if she wanted to go outside, and she gave me a look like, 'Thank God. Let's go!'"

So Suzanne took off her white coat, and they walked outdoors around the hospital. "Within minutes, she took my hand, and her parents gasped because she'd never done that with anyone before."

This yearning to go outside with her patients stemmed from intuition, which she later learned was supported by research. Simply put, nature promotes healing. Studies show that trees emit chemicals called phytoncides, which boost the immune system. Another study published in *Frontiers in Psychology* revealed that a twenty-minute experience in nature reduced levels of two stress hormones, cortisol and alpha-amylase, found in saliva. As a doctor, Suzanne now views her mission as communicating the science of nature—including the power of nature connection in cities, such as looking at clouds in the sky. She's become an advocate and practitioner of nature therapy, especially a technique known as forest bathing, developed in Japan in the 1980s to counteract the high levels of stress in Tokyo.

"I'd like to prescribe 120 to 150 minutes a week in 'nearby nature,' just as doctors now prescribe exercise to their patients."

"Even looking at images of nature or a potted plant brings some benefits, if you can believe it," she said. "I'd like to prescribe 120 to 150 minutes a week in 'nearby nature,' just as doctors now prescribe exercise to their patients."

Her journey toward integrative medicine, a more holistic form of medicine, began with two life-changing health events in her family: her son was diagnosed with autism spectrum disorder, and her husband with lung cancer.

"These are both environmentally-related conditions," she said. Growing up in Iowa, she worries about the health impacts of pesticides from the thirteen million acres of monocrops for corn. She also recognizes the importance of diversity in nature for the well-being of ecosystems.

"We have a high incidence of lymphoma and leukemia among farmers in Iowa," she said. "And we have reason to believe my late husband's cancer was influenced by radon and the chemical exposures in his work as a general contractor."

By chance, she attended a lecture on integrative medicine with the renowned Dr. Andrew Weil and applied for a fellowship at his clinic. But several times she needed to defer, as her husband was in treatment for cancer.

"Dave was in hospice, yet insisted I do the fellowship," she told me. He weaned himself from ten to three liters of oxygen to fly on the plane with her. Yet after the first week of the fellowship, her husband took a turn for the worst—and she had to face his death. Soon after, she quit her job as an OB-GYN.

"I was a shell of a human being at that point," she said. "But I knew what direction I wanted my life to go."

After Dave's death she started running and mountain biking to fulfill her adrenaline needs in ways other than delivering babies, and completed an intensive training in herbal medicine. Then she saw an article in *Oprah* magazine about forest bathing, so she tore out the pages, not realizing the technique would change her life.

While leading a workshop at a Franciscan retreat center, she decided to experiment with *Shinrin-yoku*, or forest bathing, the simple practice of walking in nature in a mindful way. After attending a six-month practicum with the Association of Nature and Forest Therapy, Suzanne realized how the techniques could apply to outdoor recreation.

She's written a book *The Outdoor Adventurer's Guide to Forest Bathing: Using Shinrin-Yoku to Hike, Bike, Paddle, and Climb Your Way to Health and Happiness*. In my classes I've also led students in forest bathing on campus, but I'd never considered expanding the practice when running trails or rafting rivers.

"When I'm working with mountain bikers, I invite them to take off their shoes and walk barefoot on the trails they ride," she said. "In fifteen minutes, they have 'aha' moments when they connect with the earth beneath them, rather than only thinking about shredding trails."

Healing people and the planet are interconnected, which might begin twenty minutes at a time.

KANSAS

FROM SOLAR PANELS TO CHARGING STATIONS:
ELECTRIFYING THE ENERGY ECONOMY

POOJA SHAH

ELECTRICAL ENGINEER

Kansas City, Kansas

The power lies in connecting the dots. That's what Pooja Shah thought one afternoon in Kansas City, when she connected an electrical bulb to a solar panel to generate energy for middle school students as a part of the Boys and Girls Club. Most kids don't understand how electricity is generated, she told me, and so she demonstrates with hands-on activities that involve STEM, the acronym for Science, Technology, Engineering, and Math. Then the students ask questions about her work as a lead electrical engineer on solar and energy storage.

"I try to help the girls see how the scientific concepts can actually allow them to make a difference in the world," she said. "If they think STEM is just about school, it's not relatable, but if we connect the dots to real-world applications, they can see the impact on the world."

She's left her own mark as a woman of color in STEM and as a mentor for young people who might never have considered a future in the energy sector. Pooja knows what it's like to walk into a meeting or enter a Zoom call as the only woman or person of color. Her advice to young people who want to address climate change through STEM is to find allies who can be mentors and cheerleaders within an organization or industry.

Pooja's focus as the lead electrical engineer for one of the largest construction and engineering companies in North America involves energy storage: how to increase the amount of renewable energy on what's called grid-scale systems. As one example, she's working on some of the largest renewable energy projects in the United States, building systems that could power 300,000 to 500,000 homes for customers.

"These large-scale projects have a lot of potential to help address climate change as we work to electrify our economy," she said. "We're looking at how to store energy when the sun is not out, and also create a more resilient grid for extreme weather conditions, including hurricanes and wildfires.

Electrifying the economy with sources of renewable energy, rather than coal or natural gas, is one step toward confronting climate change. Pooja also has worked on developing infrastructure for electric vehicles since transportation is a large source of emissions. In her work she is focused on equity, such as how to make electrified vehicles more accessible through public transit and school buses.

While her projects center on solar and emerging markets, I asked why she decided to work for a company that also builds infrastructure for the conventional oil and gas industry. How does she reconcile her values of equity and inclusion with that reality?

"I saw this position as an opportunity to make a difference," she said. "The company where I work is huge in the power sector, and this is a transition period in the energy sector. Solar and renewables are not just the right thing to do now, but the economical thing to do. I wanted to be inside a large company to accelerate that energy transition."

> "Solar and renewables are not just the right thing to do now, but the economical thing to do."

Her home in Kansas City is far from where she grew up in the western part of India, in the state of Gujarat. In that region daily temperatures were often one hundred degrees in the summer, a foreboding of more extreme temperatures to come, and coal-fired power plants were the main source of fossil fuels for electricity. While she knew she'd deeply miss her extended family in India, she moved to Syracuse, New York, to pursue her master's degree in energy engineering with a focus on renewable energy systems.

She'd never been to the Midwest when she accepted the job in Kansas City, but described herself as young and enthusiastic. "So I packed my bags and off I went! It turns out that I like the people and the work, and I've found community."

As a way to connect to and champion women in other countries, she volunteers as a mentor for a group called Global Women's Network for Energy Transition. Once a month on Zoom, Pooja consults with mentees in other parts of the world and provides technical and leadership advice on thriving in the clean energy industry.

Describing herself as an introvert who is empathic, she sees her strength as her ability to listen and empower people to find solutions together, no small feat when addressing large-scale problems. When she considers the future, she describes a concept called the energy trilemma, which includes energy security, energy equity, and energy sustainability. If we can meet energy needs while keeping energy affordable and sustainable, connecting the dots could help create a livable world.

MICHIGAN

SPEAK UP FOR CLEAN WATER:
BECAUSE IF YOU DON'T, WHO IS GOING
TO HEAR YOU?

MARI COPENY

STUDENT, ACTIVIST, & PHILANTHROPIST

Flint, Michigan

The background music played the lyrics to "I'm just a kid" by the band Simple Plan, complete with electric guitar and rocking drums. On the TikTok video, eight-year-old Mari Copeny stared into the camera with a look of steely resolve. She held a piece of white paper with the handwritten statement: "Flint, MI has been without clean water since April 24, 2014." The next image revealed thirteen-year-old Mari with a wide-open smile and a black hoodie, holding a similar sign with a pose of grounded determination. In less than a minute, the two images communicated the span of time—seven years, or half her lifetime—since she'd faced foul-smelling water in lead pipes that threated the health of thousands of children in her city.

During Black History month, a meme mapped the timeline of this young activist's life on the Instagram page of "Little Miss Flint," as she's known.

"At 6, the water in my hometown was switched.
At 8, I convinced President Obama to come to town and approved $100 million to begin fixing the water crisis.
At 12, I had given out more than one million bottles of water.

At 13, I have my own filter bringing clean water to those
   who need it.
I am Black history."

With an effervescent presence, Mari earned the moniker "Little
Miss Flint" when she wrote President Obama a letter inviting
him to see the impacts after the city of Flint switched its drinking water supply from Detroit's system
"I am Black history." to the Flint River to save money. The
discolored water with a horrific smell
resulted in lead poisoning from the dangerous and outdated
lead-based pipes, poisoning ignored by government officials.
Studies later revealed the contaminated water doubled and
even tripled the incidence of elevated lead levels in blood, a
medical disaster for children.

In response to her letter, President Obama came to Flint
and secured $100 million in funding to begin fixing the water
system, although it was the multiple lawsuits by citizens that
enforced delivery of bottled water to homes without filters,
replacement of the lead pipes, testing the water, and more.
The Michigan Civil Rights Commission ruled the crisis was
a result of systemic racism, which was clear to Mari from the
very beginning of campaigning.

After Obama's visit, Mari continued to use her growing
platform to bring resources to her community and draw attention to the environmental injustices around water and climate.
She began by raising money for bottled water and distributed
more than a million bottles, although she moved away from
single-use plastics by partnering with a company called Hydro-
viv, which produced her own water filter. While her Instagram
page was managed by her mom, Mari used social media to raise
half a million dollars to distribute these water filters.

In 2017 she was a national youth ambassador to the Wom-
en's March and the National Climate March, and has expanded

her reach through media such as Cartoon Network, innovating ways to share the story of climate action to children.

"Cartoon Network surprised me and made me a Power Puff Girl," she said, with her own animated character, "Global Water Warrior," with big brown eyes and a flower headband. This platform seems appropriate, since watching anime is one of the activities she enjoys when taking a break from activism. Before elections, Mari used her voice to encourage others to vote, even though she wasn't able to yet.

On primary day in Flint, she wore her "We vote next" T-shirt, a sign that the voice of youth will be at the ballot box soon. Standing in line with her mom, Mari told her, "It's not fair I have to wait until I'm 18 to vote." A poll worker overheard her and said she shouldn't be rushing to vote. But that person didn't realize Mari intends to run for President in 2044, and until then, she's going to keep speaking up.

In 2020 the state of Michigan came to a settlement with victims to pay $600 million to those who suffered damages in the Flint water crisis, most notably children exposed to lead, when the EPA reported there is no safe level of lead in water. In 2021 the former governor Rick Synder was charged with willful neglect of duty in response to the Flint water crisis, although the charge was ruled a misdemeanor with no more than a $1,000 fine or one year in jail.

In response to these charges, Mari wrote on Instagram: "It's 2021, Michigan finally charged former governor Synder for his role in poisoning my entire city. Flint is not fixed."

And Mari isn't waiting for action from others, most notably adults. She's aiming for $1 million in funding for water filters and charting her course—as an animated hero and a real-life leader for our times.

MINNESOTA

WATER IS LIFE:
THE CLIMATE MOVEMENT NEEDS
INDIGENOUS VALUES AND VOICES

TARA HOUSKA

ATTORNEY AND ENVIRONMENTAL AND INDIGENOUS RIGHTS ACTIVIST

Couchiching First Nation, Northern Minnesota

Sometimes a person can't help but make a few observations. That's exactly what Tara Houska did when thousands of domestic terrorists, mostly white men, stormed the US Capitol building on January 6, 2021, with little organized resistance and only thirteen immediate arrests.

She contrasted it with her experiences as an Indigenous woman from the Couchiching First Nation: "A few weeks ago, 22 of us were arrested here on my people's homelands standing in front of machines destroying our homelands for an oil pipeline," she wrote on Twitter. Then she listed her experiences defending Native land and drinking water from fossil fuel destruction: rubber bullets, mace, arrest, tear gas, strip searches, and even being locked up in a dog kennel.

As an Ojibwe attorney and activist in northern Minnesota, she's also known by her Indigenous name Zhaabowekwe as she fights pipelines that threaten the health and future of her tribe. And she sees how Indigenous values can redefine our relationship to the land and water on which we depend, especially since Indigenous lands hold 80 percent of the earth's global biodiversity.

"A river is a living being that gives life," she said in an interview with The Years Project. "If you don't drink, you will die." Native water protectors and land defenders are trying to halt the tar sands pipelines in northern Minnesota, like the Enbridge Line 3, which will go through the wild rice fields of Ojibwe territory and end at Lake Superior in Michigan, threatening the health of their food, water, economy, and culture.

Her role as a leader with a global platform fighting multinational energy companies began in the tiny town of Ranier, Minnesota. An avid reader who biked eleven miles to the library, she became the first person in her family to graduate from college and then law school at the University of Minnesota. Working for a private firm in Washington, DC, she spent days on the Hill and advocated for the removal of Native symbols as mascots, when such a request was often met with resistance.

It was while working on the Bernie Sanders 2016 presidential campaign that she met Indigenous activist Winona LaDuke and joined her organization, Honor the Earth, as an attorney. From there, she went to Standing Rock Sioux Reservation, where thousands were fighting the Dakota Access Pipeline, recognizing that Indigenous people are impacted first and worst by the fossil fuel industry.

At Standing Rock she spent six months protecting sacred lands and water. During that time, she was arrested and put into a dog kennel with others, and it was also at Standing Rock where she felt the collective energy of connecting with the Earth through Indigenous values.

In northern Minnesota, Tara has lived in a shed without electricity and running water on seventy acres among the Giniw Collective, founded to foster relationships between people and the land. Yet she seems as comfortable in the forest as on the 30th floor of a New York skyscraper, advocating for bank executives to divest from fossil fuels—and suing them as well.

On social media I've followed her fight against the Line 3 tar sands pipeline and Enbridge, the largest fossil fuel infrastructure company in North America. I watched videos of water protectors, land defenders, and allies bundled up as they chanted, sang, and prayed around a giant wooden tripod. Twenty feet in the air, a young woman was suspended between the poles, blocking the entrance to Enbridge's US tar sands terminal.

Their signs read: "Stop Line 3." Drums beat. The ground was frozen solid.

The police arrived and began to saw the legs off the tripod, causing the pine poles to wobble. "I see red as I try to speak clearly and calmly to reach these people who would directly, intentionally, harm a human being for the profit margin of an oil company," she later wrote in *All We Can Save*. The young woman let herself down and was taken to jail, and later bailed out by her allies.

While thousands have protested in northern Minnesota, the lawsuits continue for and against these tar sands pipelines. In the meantime, banking institutions are divesting from fossil fuels worldwide. The resistance is making a difference toward a future fueled by renewable energy, but Tara cautions that changes in the climate movement will require vastly different systems—not ones modeled after systems of inequity, based on individualism and capitalism: "Quite simply, I don't believe we will solar panel or vote our way out of this crisis without radically reframing our connection with our Mother."

> "Quite simply, I don't believe we will solar panel or vote our way out of this crisis without radically reframing our connection with our Mother."

MISSOURI

HOW MONEY CAN CHANGE THE WORLD:
IMPACT INVESTING IN THE HEARTLAND

EMILY LECUYER

IMPACT INVESTOR

Kansas City, Missouri

When many people think of high-stakes finances, they think of Wall Street. When Emily Lecuyer considers how money matters, she thinks of America's heartland. As the managing director of equity^2 LLC, she's part of an impact investing firm focused on investment opportunities in historically excluded communities in the Kansas City region in Missouri. In a sense, her work brings Wall Street to Main Street.

"We were presenting the opportunity to invest in a local juice company to our Community Advisory Board, and one board member suggested we track local purchasing of produce—in addition to the social and economic metrics like number of individuals hired from low-income backgrounds and number of products sold," she told me. "Now we are also measuring how the company impacts carbon emissions by buying local."

By working with local farms that quantify their carbon sequestration, she hopes to increase the accuracy of this tracking over time. "All investment has impact, whether it's positive or negative," Emily said. "We look for shared prosperity—financial returns but also measurable social and environmental returns."

To that end, equity^2 invests in BIPOC founders and purpose-driven businesses that create quality jobs accessible for residents of surrounding low-income communities. Emily's team is also launching a Community Investment Trust (CIT), where for as little as $10 a month, residents of gentrifying neighborhoods can invest in real estate in their own backyards. It's a way to build a more equitable regional economy.

"If early in my career, someone had told me I'd be working in finance, I'd be a mix of embarrassed and shocked," she said. "But I didn't know what I know now."

"We're trying to do more than just make money," Emily said. Two years ago while pregnant with her oldest child, Aurelian, she launched the firm. She's now a mom to a toddler named Truman too.

"There are big challenges in our world—climate change, food insecurity, health care, racial inequity—and impact investing can help to shape local communities for the better."

According to the Global Impact Investing Network, impact investing now accounts for more than $700 billion worldwide, even in my town of Asheville, North Carolina, home to a community equity fund for BIPOC business owners. While that's a miniscule part of overall global capital markets, it's a model for building climate resilience on a regional scale.

As a student in high school, Emily discovered she could make a difference amidst the enormity of the world's challenges by working in food and agriculture. She got involved in farmworker rights, studied sociology and anthropology in college, and then worked alongside farmers in the Philippines as a Peace Corps volunteer.

"Candidly, money is really powerful, and if I wanted to do something good in the world, I felt I had to understand how capital flows. How can you change something if you don't understand how it works?"

After the Peace Corps, while working on a small-scale diversified farm and then managing a farm business incubator for resettled refugees, she kept asking questions about money. "I felt like access to capital was this challenge that I didn't understand," she said. "Candidly, money is really powerful, and if I wanted to do something good in the world, I felt I had to understand how capital flows. How can you change something if you don't understand how it works?"

So she started out in a community bank, learning about financing small businesses and challenging historical obstacles, such as redlining, the discriminatory practice by which banks and insurance companies limit access to loans or other services based on race and ethnicity. She's applied that knowledge to equity[2], which is owned by the Community Development Financial Institution AltCap, and the neighborhood grant-making nonprofit Community Capital Fund. The firm's first $5 million fund focuses on businesses that need equity investments of $75,000 to $300,000, an amount that can make a difference for growing companies without access to resources from family and friends.

In 2021 equity[2] invested $250,000 to purchase the Marlborough School, an abandoned but stately red-brick structure on the corner of 75th and Troost Street. In collaboration with Marlborough Community Land Trust and Marlborough Community Coalition, equity[2] is redeveloping the site by making green infrastructure a priority. Through the Community Investment Trust, current residents will ultimately be the property owners.

"It is so much easier for people to do what they know—to either think with their philanthropy hat or with their profit-maximizing hat," she said. "We're asking them to do something in between. The Show Me state may be slow to catch on, but once we do, it's my hope we'll be leading the way." Emily's role as an impact investor—on the ground in Kansas City—ensures money matters by addressing community needs in the heart of the country.

NEBRASKA

WHEN CLAY MEETS CLIMATE:
AN ARTIST SCULPTS OUR RELATIONSHIP TO WATER

JESS BENJAMIN

CERAMIC SCULPTOR

Omaha, Nebraska

Pounding, mixing, drying, compressing, pushing, moving, manipulating, carving. These are the verbs of her vocation. For sculptor Jess Benjamin, creating art is a physical act in body and imagination, addressing water and drought in our changing climate.

As an artist who grew up on a ranch and farm in Cozad, Nebraska, Jess drives a Chevy Silverado truck and wears a feed-lot cap, a practical accessory to hold back her hair while working and a throwback from her childhood. In her 4,300-square-foot studio and gallery in the city of Omaha, she creates large-scale ceramic artwork focused on water usage in the Great Plains region and the nation. As Jess works, a sleek black cat, a ginger-colored dog, and her seven-year-old daughter are frequent companions looking up to her from the studio floor.

Several feet above that floor, after climbing the steps of a footstool, she arrives to her workstation, planting her feet on narrow pieces of wood atop stacked concrete cinderblocks. There, she can reach the eight-foot sculpture depicting the massive intake towers that control water at the Hoover Dam, each weighing 600 pounds in the studio. She's shaped these towers of clay to resemble concrete and rebar, revealing disintegration due to drought.

The Hoover Dam's waterline is currently at a historic low.

"In the 1930s we built dams to store water, but the system doesn't work if it's not snowing or raining enough," she said.

She's sculpted jackstones, the tetrahedrons on the surface of dams, which remain underwater until exposed during drought, such as at Kingsley Dam at Lake McConaughy near Ogallala, Nebraska. Indeed, the Ogallala Aquifer is the largest underground aquifer in the world, and she sees water as a resource humans will need to conserve, pipe, transport, and value as the climate changes.

Her art is deeply rooted in place: agriculture, aquifers, family, the Great Plains. "An artist's job is to make things beautiful that aren't," she said. "Historically, artists were paid to make images. Today, I need to document this time period. I have both rural and urban backgrounds—it's a unique perspective on the value of water."

Clay is her medium, visible in each of her pieces: "Clay needs water to be built, and you have to add fire to take the water out," she said.

Since she was four years old, Jess has experimented with clay. She played in the irrigation ditches of her family's farm and ranch, and when the topsoil eroded, the clay appeared, which she'd use to build dams and divert streams.

"Making art about water is also a way for me to stay connected with my family," she said. "I know what the weather and water patterns have been in my lifetime, but my father recalls the weather before the 1960s. My family runs a feedlot, farms, ranches, grows corn, harvests, feeds the corn to cattle, and watches run-off water flow into irrigation systems."

Now these currents have all come to rest in the heartland.

"Have you ever been to Nebraska?" she asked me, out of the blue.

"Yes, it was beautiful, but my visit was three decades ago," I admitted. My memory of the state comes from photos of

Sandhill cranes and wide-open plains. Acting as a long-distance nature guide, Jess described the beauty of the Sandhills, the grass-covered stabilized sand dunes that cover one quarter of the state.

"What happens in the middle affects the rest of the country," she said. "But we don't get the same news coverage. When drought affects California, it's in the news, but the media isn't always paying attention to Nebraska."

> "What happens in the middle affects the rest of the country."

As a farmer and rancher's daughter, she trained her imagination on the 100th Meridian, the line that runs down the middle of the country. As a sophomore in college, she discovered ceramics as a medium well-suited for her tactile experience with this sense of place: earth, water, and fire, and of course air.

In her gallery one of her pieces depicts giant faucets, buckets, and spigots, a visual prompt to consider water consumption in our own lives. Another reveals a color-coded map of the Ogallala Aquifer, which she created by sculpting water molecules. Her work draws on scientific data as well as observations and interviews with those who use water for their livelihood, especially farmers. And when she's not creating art, she supports the work of others in her role as the director of the Lied Art Gallery at Creighton University in Omaha.

"If you consider civilizations throughout history—the Egyptians, Incas, Aztecs, Romans—how they moved, transported, conserved, and accessed water was critical to their survival and eventual collapse," she said. "I am taking a positive outlook, but using my art to ask questions and document our times."

NORTH DAKOTA

WOMEN ARE THE WORLD:
INDIGENOUS CLIMATE LEADERS, STRONGER TOGETHER

KANDI MOSSETT WHITE

INDIGENOUS AND ENVIRONMENTAL RIGHTS ACTIVIST

Fort Berthold Indian Reservation, North Dakota

They passed the baby around the circle, as mothers have done for hundreds and thousands of years. It could have been any gathering of women, cooing at the swaddled infant, swooning with words of endearment. But these women gathered were climate defenders from all over the world—Kenya, Israel, Sudan, Columbia, Nepal, Nicaragua, and Guatemala. They'd arrived on Fort Berthold Indian Reservation, not only to greet this new child, but to learn from his mother, Kandi Mossett White, about her organizing work as an Indigenous woman living on land desecrated by the oil and gas industry.

In years past they'd met as a delegation in New York City for the United Nations Permanent Forum on Indigenous Issues. But in 2019 they came to this tribal homeland in North Dakota on a journey organized by the organization MADRE and the Indigenous Environmental Network, for whom Kandi works as the director of Native energy and climate campaign.

"Look to the left, look to the right," said Kandi. "You see rig after rig, flare after flare. It's just hard to know everybody's breathing this in every day."

During their time together, the group shared stories of their activism and the ways they confronted the devastation of climate change, war, and extractive industries with creative, collaborative strategy. But exposure often came at a cost. Some had been jailed for their activism. Others suffered from cancer caused by the cumulative impacts of toxic waste. At night during their visit in North Dakota, they couldn't see the stars due to the light of the natural gas flares.

As their host, Kandi led the group to sites such as a 2014 saltwater spill—or brine spill—a toxic byproduct from fracking operations, which killed trees in the area with salt leaching into the land and into groundwater. Kandi introduced herself as a Mandan, Hidatsa, Arikara woman who grew up in a small rural community, homeland of these three Affiliated Tribes in North Dakota. Her Hidatsa name is Eagle Woman. As a child, she played outside all the time, swimming, fishing, picking chokeberries, and exploring the Badlands.

"I thought North Dakota was the best place on earth," she wrote for the *Earth Island Journal*. But as a young person, she didn't see anything unusual about the power plants and uranium mines nearby—or the growing number of diagnoses of cancer, diabetes, and asthma in her family.

"I never knew that it wasn't normal for like ten people in your family to have cancer," Kandi said, "until I left here and went to college and started meeting people who never knew anybody who had cancer. And I was like, 'Really?'"

As a twenty-year-old, she was diagnosed with a stage 4 sarcoma tumor, for which she had three surgeries, forgoing chemotherapy and radiation due to concerns about possible impacts on her reproductive health. Surviving cancer changed her perspective as she became more aware of the links between cancer rates in her community and nearby oil and gas industries. A recent review of the literature suggests health impacts due to upstream oil extraction in the United States include cancer, liver damage, immunodeficiency and neurological symptoms, with additional adverse impacts on soil, air, and water quality.

After joining a group that blocked a tar sands oil refinery, she began working for the Indigenous Environmental Network in 2007. About this time, fracking came to Fort Berthold, where the women actually opposed decisions to proceed with hydraulic fracturing made by the all-male Tribal Council.

"It has often been said that the rape of Mother Earth is connected to the rape of women," Kandi wrote. With the growth of fracking came the influx of male workers. More violence against women was reported, as well as an increase in sex trafficking in the area.

> "It has often been said that the rape of Mother Earth is connected to the rape of women."

She sees women as the grassroots leaders in the movement, the "keepers of the water and carriers of the next generation." That power is what came with her to Standing Rock in 2016, as she stood up against the Dakota Access Pipeline, where she joined thousands of water protectors working toward a just transition for the future.

At the global gathering where she brought her infant son, she held hands in a circle with the women who'd traveled from across the world to strategize together. The Indigenous women around her formed the backbone of the work to protect Mother Earth.

To that end, Kandi is building a global network around the world. Her message of solidarity, rather than isolation, carries the seeds of her vision—which is more than just fighting against extraction. She envisions a world with large-scale community gardens, energy from the sun and the wind, and jobs for all in her community.

OHIO

TESTIFY YOUR TRUTH:
A FRACKING REFUGEE SPEAKS UP IN APPALACHIA

JILL ANTARES HUNKLER

COMMUNITY ORGANIZER

Barnesville, Ohio

"I consider myself a fracking refugee," said Jill Antares Hunkler on Earth Day, April 22, 2021. She wasn't making small talk at a tree-planting ceremony or a neighborhood cleanup. Instead she joined activists like Tara Houska and Greta Thunberg to testify virtually before the House sub-committee on the environment. "My intention is to share this truth," she said. "Continuing to subsidize the fossil fuel industry will not only perpetuate the climate crisis, but the plastics pollution, environmental justice, and public health crises as well."

As a seventh-generation resident of southeastern Ohio, Jill described her home in Appalachia as "occupied territory" because of the influx of the fossil fuel industry from the frack-ing boom. Her former house is now surrounded by oil and gas infrastructure, and the air and ground are filled with the pollu-tion from a compressor station, seventy-eight fracking wells, a transfer station, and an interstate pipeline—all industries with-in a five-mile radius of her home.

Given the harm caused by extracting natural gas from shale, hydraulic fracturing has overtaken the farmland and country roads in Belmont County, the most fracked county in the state with 595 producing wells. Her home was situated at the headwaters of the Captina Creek Watershed in a hollow with Slope Creek running through her yard.

"The air pollution emanating from these facilities contains volatile organic compounds . . . that are heavier than air and hover in the hollows," she said. When she began to exhibit symptoms of nausea, vertigo, nose, ear, and throat irritation, she collaborated with scientists to develop low-cost monitoring systems, organize public forums, and document the impacts of fracking. Ultimately, she had to sell the property in 2020: "I was so sick. The health impacts were unbearable," she told me. "I had to give it up. When I think about having to move, it just makes me sad."

"I consider myself a fracking refugee."

Describing herself as shy and reserved, Jill grew up in Barnesville, Ohio with extended family and cousins who were like siblings. But she also lived all over the country with her mom and later traveled the world as an adult. Thirteen years ago she returned to stay in an Airstream on family property with her daughter and hand build a new home from recycled materials, using bricks from her great-aunt's house.

She is most comfortable in nature, where she practices healing rituals and ceremonies to protect Mother Earth, inspired by Indigenous cultures and the diverse religious traditions of the world. In her home she surrounds herself with sacred objects of meaning: a painting of Bob Marley, a ceremonial staff, and baskets from Kenya.

But her heart is in her hometown of Barnesville, a village of 4,000 people, where 80 percent of the landowners signed leases for oil and drilling on their property. "Hundreds of people were lined up around the high school to sign up," she said.

While 14 percent of the population lives below the federal poverty line, the promise of payment from the leases came with costs. Jill's activism was spurred by events such as a 2018 fracking well blowout in Belmont County, producing one of the largest methane leaks in US history as residents were forced to evacuate. The amount of greenhouse gas methane released was 60,000 tons, more than some countries emit in an entire year. Now, with Jill's assistance, researchers like Nicole Deziel from the Yale School of Public Health are studying long-term consequences of fracking in Belmont County as well as collecting growing evidence of negative health impacts.

During the testimony to congressional leaders, Representative Ro Khanna from California said the United States is the world's second largest producer of fossil fuels, and the second largest subsidizer of fossil fuels. Youth activist Greta Thunberg then asked the Congressional leaders: "How long do you think you can continue ignoring the climate crisis without being held accountable?" Indeed, studies from Penn Future found that fossil fuel subsidies—money paid to oil and gas companies—sustain global greenhouse gas emissions at levels 28 percent higher than the market would.

While the oil and gas industry points to jobs as an outcome, Jill shared the on-the-ground reality in Appalachia: "In the years since the fracking boom began, Belmont and other eastern Ohio gas-producing counties haven't gained jobs because of fracking. In fact, they have lost more than 6,500 jobs, according to the Bureau of Economic Analysis and the Bureau of Labor Statistics."

She sees a future fueled by renewable economies, regenerative agriculture, and strong infrastructure, rather than a "sacrifice zone that is resource-cursed." As a voice in Appalachia for those living with extractive industries in their backyards, Jill won't give up on her vision to bring clean air, water, and soil to her home.

SOUTH DAKOTA

HARNESSING PAIN INTO ACTION:
YOUTH DEFENDERS OF WATER AND LAND

JASILYN CHARGER

INDIGENOUS ENVIRONMENTAL PROTECTOR AND LAND DEFENDER

Cheyenne River Reservation, South Dakota

Sometimes people make a difference because they respond to a call and just keep doing the next right thing. That's what happened when Jasilyn Charger heard a message on the radio from members of the Standing Rock Sioux Tribe fighting the Dakota Access Pipeline. Help was needed from Indigenous activists who'd gained experience marching against the Keystone XL pipeline. When young people such as Jasilyn heard the call, they did the next right thing: "We were like, 'Alright!' We brought our youth movement."

Upon arrival, Jasilyn learned they were the only ones who'd answered the call at the time. One Mind Youth Movement had formed as a support for teens facing substance abuse, suicide, and poverty—legacies of systemic racism and colonization on the Cheyenne River Reservation in South Dakota. Those legacies of harm were often addressed through the group's activism to confront change on both personal and community levels.

At nineteen, Jasilyn lost their best friend, one of eight other youth who'd died by suicide that summer on the reservation. Native American youth are one and a half times more likely to kill themselves than the national average. These horrific losses became fuel for next steps that would inspire not only Jasilyn but also youth across the country.

As Jasilyn describes it, their own challenge with addiction was the first battle they ever won. "I became a water protector and earth defender because I was protecting myself," they said in a video called "Activist of the Land."

In 2016 after Jasilyn arrived at the Standing Rock Reservation, they intended to stay for a few days, or maybe a week, and ended up staying a year. Most images in the media from Standing Rock portrayed a vast expanse of teepees, flags, drums, and thousands of people galvanized in this legal and spiritual fight for sacred land and water. But somebody had to set up camp first: Jasilyn was one of the first five people to camp after showing up to support the Standing Rock Sioux Tribe.

"We shared how to better communicate with their youth," they write in *How We Go Home: Voices from Indigenous North America*. "You know, not just to say, 'You have to be here,' but 'Can you help out?' If you're an artist, can you make a banner?" Youth also used their social-media savvy in support of the water and land threatened by the Dakota Access Pipeline.

Living in a teepee wasn't easy without access to electricity or heat for Jasilyn and the other youth: "But then we turned a spiritual corner," Jasilyn writes. "We thought, we're here. This is like a prayer. And it kind of reconnected all of us to our heritage because that's how we used to live."

But they didn't grow up thinking about themselves as an activist, but rather as a shy child focused on survival. In fact, when their mother was unable to care for her children, Jasilyn rotated between the foster care system, the homes of family members, and a youth psychiatric facility, separated from their mom and twin sister until the age of eighteen. They were tragically raped on that eighteenth birthday, as they wrote in a longer autobiographical piece. In those writings, Jasilyn shares these intersecting traumas with a matter-of-fact, stream-of-consciousness rhythm and witness. With a bandana around their forehead, framing jet-black hair, Jasilyn projects a grounded confidence when they speak, now as cofounder of the International Indigenous Youth Council.

Staging nonviolent actions became a way for Jasilyn to use their organizing skills to impact policy. While at Standing Rock, Jasilyn helped to coordinate a 2,000-mile run from North Dakota to Washington, DC, in order to present a petition to the US Army Corps of Engineers. Representing all Sioux reservations, young people ran in stages and then handed off the role of runner in a cross-country journey raising awareness of water rights across the states, while many in the media followed the runners.

Yet during their months at Standing Rock, despite these acts of nonviolence, water protectors and land defenders like Jasilyn faced police in SWAT gear with rubber bullets and helicopters as they were later forced off the land.

> **"We took the beatings, we took the mace. It was like Armageddon, like the world had ended."**

"We took the beatings, we took the mace," Jasilyn writes in *How We Go Home*. "It was like Armageddon, like the world had ended."

In 2020 the courts ordered the Dakota Access Pipeline shutdown pending environmental review, and the US Supreme Court later rejected an appeal from the pipeline operator to avoid the review. In addition, President Biden halted the Keystone XL pipeline on his first day in office. What we know for sure is that countless climate activists found their collective footing in the movement at Standing Rock, which started when a small group of people, including Jasilyn, decided to do the next right thing.

WISCONSIN

AMPLIFY DIVERSE VOICES:
BUILDING ON A FAMILY TRADITION
OF EARTH DAY FOR ALL

TIA NELSON

PUBLIC SERVANT AND ENVIRONMENTAL ACTIVIST

Madison, Wisconsin

Tia Nelson found her voice early and hasn't been afraid to use it—even when it meant resigning from her job as the executive secretary of Wisconsin's public lands agency. When the Republican-held board—in a two-to-one vote—banned the agency from talking about climate change in 2015, she walked. She'd come to the public lands agency from The Nature Conservancy, an organization representing over twenty-five countries on climate policy and forest conservation. By the time she came back home to Wisconsin, she'd spent eleven years putting that experience to work—only to be told she couldn't discuss climate, or later, advocate for climate policies. Her upbringing taught her to persist in spite of challenges. So that's what she did.

Tia was only two years old when her father, Gaylord Nelson, was elected governor of Wisconsin. Called the "conservation governor" for his environmental policies, he drew support from Republicans and Democrats alike. But when he was elected to the Senate in 1963, and the family moved to Washington, DC, he faced obstacles garnering support for environmental protections.

Something changed for him in the fall of 1969, when he read an article about the impact of the anti-war teach-ins on college campuses on national discourse about the Vietnam

War. Inspired, he called for a national teach-in about the environment. He enlisted a Republican Congressman, Pete McCloskey, to cochair his Earth Day advisory committee, which was formed during a time of national tensions around the Vietnam War, racial inequality, student protests, and unrest heightened by the assassinations of Martin Luther King, Jr. and Robert Kennedy. Despite the divisions, Tia recalled two million people gathered together for that first Earth Day in 1970, more than fifty years ago.

"What many people don't know is how long and challenging a journey it was for my father to get to that place of success and how many setbacks and disappointments he experienced along the way," she said in an interview with Citizen's Climate Lobby. She described her sense of duty in leaning into his legacy, which for her meant meeting people where they were and bringing them into the movement. From her father, she learned the value of amplifying diverse voices and finding common ground, even when it might not be obvious to others.

On that first Earth Day as a multigenerational, bipartisan event, Tia was thirteen years old, participating in cleanups to mark the day. The years that followed saw passage of the Clean Water Act and the Clean Air Act with bipartisan support, but Tia said her father would have been troubled by the denialism of science so polarizing in the halls of Congress today.

To continue her work and honor her father's legacy, Tia helped to produce a short film "When the Earth Moves," which showcases historical footage from that first Earth Day, with interviews from two voices for the climate on opposite ends of the political spectrum: Varshini Prakash, cofounder of the Sunrise Movement, and Bob Inglis, former climate skeptic and Republican Congressman from South Carolina.

"Growing up as his daughter, it was mostly a great privilege, but also I felt a heavy sense of duty to public service and to making a difference with my life," she said in the film, which begins with a speech her father gave on the eve of the first Earth

Day. I imagine those words would speak to my students today: "Our goal is an environment of decency, quality, and mutual respect for all human beings and all other living creatures," he said. "Our goal is a decent environment in its broadest and deepest sense. And it will require a long, sustained political, moral, ethical, and financial commitment far beyond any commitment ever made by any society in the history of man."

> "Growing up as his daughter, it was mostly a great privilege, but also I felt a heavy sense of duty to public service and to making a difference with my life."

Of course, my students would amend the language to include the "history of all humans," but they are working, much like Tia, in hopes that Earth Day is not just a festival on a field but a celebration of sustained work for all humans and other living creatures.

On Earth Day 2021, history came full circle as President Joe Biden chose this day as "the launching pad for our nation's recommitment to international action on climate change," Tia reflected. She recalled how her father welcomed twenty-nine-year-old Joe Biden to the Senate and comforted the distraught young senator a month later when Biden's wife and daughter died. Tia's father encouraged him to try the job for six months and invited him into their circles.

"He would be moved to see that the heartbroken young man he helped recover from despair is carrying his legacy forward," she writes. While President Biden's climate legacy is still up for debate, Tia continues to honor her father's foundation with her own life's work.

WEST

ALASKA

FOLLOW THE MONEY:
PRESSURING BANKS TO PROTECT CLIMATE,
CARIBOU, AND CULTURE

BERNADETTE DEMIENTIEFF

EXECUTIVE DIRECTOR

Fort Yukon, Alaska

It's not a stretch to describe the fight for the Arctic National Wildlife Refuge as a David and Goliath battle, pitting banks, the fossil fuel industry, and politicians against the very people who depend on the caribou of the Coastal Plain of the refuge. Bernadette Demientieff would know: she's fighting for the survival of this sacred land, the calving ground for 200,000 caribou, vital to the food, tools, clothing and spiritual guidance of the Gwich'in Nation.

One of her strategies for advocacy was to travel from Fort Yukon, Alaska, to New York City to sit in a conference room on the 43rd floor of a corporate bank's headquarters and speak about protecting the Coastal Plain of the Arctic National Wildlife Refuge. She brought the story of the Gwich'in Nation to those who might fund proposed oil and gas development on Indigenous sacred land: JPMorgan Chase, Barclays, Goldman Sachs, Bank of America.

The renewed threats to the land were imminent. In 2017 Congress passed a tax bill that included a provision to open the Arctic Refuge for drilling. This move threatened the calving grounds of the Porcupine Caribou herds on which the Gwich'in depend for their survival and an Indigenous way of life that dates back 20,000 years.

After years of lobbying Congress, the Gwich'in Steering Committee, with Bernadette as its executive director, shifted tactics to target the banks providing funding to the fossil fuel industry. When talking with lawmakers didn't work, they approached their funding sources, the names on our credit cards and bank statements.

"We didn't go looking for a fight," Bernadette said in an interview with the Sierra Club. "But we are calling on banks to update their policies and rule out financing drilling on the Refuge."

The campaign involved one hundred environmental and Indigenous rights groups and investors who documented the risks of investing in fossil fuels on these lands due to the ecological impacts, the human rights violations, and the global shift toward renewable energy.

But the spiritual connection to the caribou herds, the vastness of the land, and devotion to her children are what drive Bernadette's persistence in the fight she wasn't looking for. The Arctic National Wildlife Refuge is home to forty-five species of land and marine mammals, including the caribou herds that migrate 3,000 miles a year but depend on the Coastal Plain for birthing and calving 40,000 young each year in "the sacred place where life begins."

Reflecting on her own life, she says, "I'm not the same person I was when I started this work." When her brother died by suicide in 2006, she began drinking. "I lost it a little bit," she said. "But I'm doing this now for my children."

As a mother and grandmother, she follows the spiritual direction of her elders to protect the Coastal Plain, a sacred part of the refuge. "Our animals are our survival. Our land, our clean water, our clean air—that is our survival," she said.

The 9,000 Gwich'in people live in thirteen villages along the migration route and depend on the herds for their food security. But the caribou are adapted to cooler temperatures and impacted by global warming. Populations of one of the herds in developed areas have decreased by 50 percent in the last three years, she said. "So they can't tell us our food security will not be affected when we see otherwise," she told me.

The threat to her homeland had never been greater than when the Trump administration stated its plans to open up drilling before he left office. But in 2020, Bank of America and every major corporate bank in this country updated its policies to prohibit funding oil and gas drilling in the Arctic National Wildlife Refuge, with 70 percent of the US population unwilling to support fossil fuel extraction along the Coastal Plain.

Even so, the Bureau of Land Management moved to sell oil rights on the land in the last weeks of the Trump administration. The sale ended with only three bidders, and nearly half of the leases had no bids at all. As *NPR* reported, the forty-year battle over drilling on the Coastal Plain ended in large part due to this campaign combined with low oil prices. The next step for the Gwich'in and their allies is to gain permanent protection for the land,

"Our ways of life are not for sale, and our identity is not up for negotiation."

helped by the hold on oil and gas drilling in the Arctic signed by President Joe Biden hours after his inauguration. Over a year later, Biden suspended all drilling leases in the refuge.

Looking forward, Bernadette recalls the early years when she didn't know how to write a grant or run a nonprofit but felt a strong sense of justice. She's always been an ardent believer in the power of community and prayer, even as she faced yet another profound loss, when her son was murdered.

"I miss his smiles, his voice, his hugs," she told me, tearing up as she spoke to me from a a nearby gas station known for its hot apple fritters and coffee—and good cell reception. "This work is hard. But his life is connected to this place." As she so often says: "Our ways of life are not for sale, and our identity is not up for negotiation."

ARIZONA

DROUGHT MAKES A DIFFERENCE:
USING SCIENTIFIC RESEARCH TO UNDERSTAND
ACCESS TO WATER

DIANA LIVERMAN

GEOGRAPHER

Tucson, Arizona

"Can we be the first generation that can put an end to poverty and the last generation to put an end to climate change?"

As a part of a Nobel-prize winning team, Dr. Diana Liverman posed this question to the Nobel Conference in 2019. She was quoting former Secretary of the United Nations Ban Ki-Moon to show the critical intersections between poverty, sustainable development, and the climate crisis that have been the focus of her work at the University of Arizona.

While young activists have helped to place the climate emergency in the headlines, Diana has been one of the early climate researchers during a career spanning forty years. News footage from 1997 shows her as the only woman among a panel of men including President Bill Clinton and Vice President Al Gore in a White House conference about global climate change. Since 1995, she's been a coauthor of the Intergovernmental Panel on Climate Change (IPCC), the report published every six years by leading climate scientists from around the world.

A geographer focused on climate justice, she's most interested in how adaptations to climate affect under-resourced communities. As an undergraduate in college, she realized that vulnerability mattered—in significant ways—in terms of community responses to climate impacts.

"The impact of (less) rainfall was fine if you were well-off and had access to water," she said in an interview with *High Country News*. "But if you were poor with no land and no water, then you were really going to suffer." This reality holds true for a farmer in the western United States as well as one in West Africa. Her graduate students have completed field work assessing how some adaptations to climate benefit men and not women. For example, adaptation funding for irrigation projects won't help women if they don't own land. Her research encourages governments and nonprofits around the world to consider the outcomes of adaptation, so these measures help all people, rather than creating even greater inequality.

When the IPCC Special Report on the impact of 1.5 degrees Celsius came out in 2016, she shared the scientists' assessment: "Climate change is already here. Every bit of warming matters. We have to act soon and make deep cuts in emissions to get to 50 percent by 2030 and net zero emissions by 2050." With this report, however, the media fixated on 2030 as an apocalyptic date when the world might end.

"The story was, 'We've got 12 years before it's the end of the world,'" she said. "And that was so not the message we were trying to convey. I was quite upset about young people just feeling paralyzed by that message." In contrast to that soundbite, the full report's message was loud and clear: if we delay taking global action, the world will be warmer, and it will be harder to reduce emissions and impacts.

Her life as a global climate leader began in Ghana, West Africa, where she was born since her father worked for the British government developing dams. As a young person, she had a keen interest in water and the environment and the outcomes of development on communities.

While she was raised in the UK, she pursued graduate school in the United States and Canada and taught at the University of Arizona, where she mentored more than seventy graduate students, who are now working all over the world. As a woman leading on climate, she's watched the early days of the IPCC report and experienced sexism within these circles of scientists.

"The first IPCC report (in 1990) had a handful of women," she told the *Pacific Standard*. "—I know they overcame tremendous barriers."

By the 6th IPCC cycle, 27 percent of the physical sciences group were women. With a graduate student, Diana conducted a survey of IPCC scientists, and one-third of respondents said childcare and family responsibilities prevented their full participation in this unpaid work that happens every six years on top of existing jobs.

> "The first IPCC report (in 1990) had a handful of women—I know they overcame tremendous barriers."

In 2020 the IPCC adopted a Gender Policy and Implementation Plan to enhance gender equality in their process, even as they study the impact of climate on women. That same year Diana was inducted into the National Academy of Arts and Sciences, a lifetime honor. She brought her teaching career to a close, committing herself to full-time research. When youth protest in the streets for climate justice, scientists like Diana have worked for decades, providing the data that drives the cause.

CALIFORNIA

WORDS MOVE US:
POETRY INSPIRES AND MOBILIZES
CLIMATE JUSTICE

AMANDA GORMAN

POET

Los Angeles, California

Her jet-black braids under a fire-engine-red beret, Amanda Gorman sat on an orange couch and glanced upward. With deep brown eyes rimmed in iridescent eye shadow, she seemed to look out to see how the future would unfold. Moving her fingers toward her face, as if to take a picture, she used her hands to mimic the orbit of the earth, making an arc above her head as she described "a blue orb hovering over the moon's gray horizon . . . It was our world's first glance at itself, our first chance to see a shared reality, this floating body we call home."

By opening her poem "Earthrise" with the mission of Apollo 8 blasting off on Christmas Eve in 1968, this youth poet laureate brought us to that first snapshot of the Earth and then led us to the crisis of justice facing our home: "For it is the obscure, the oppressed, the poor, who when the disaster is declared done, still suffer more than anyone."

In fact, Amanda wrote "Earthrise" in 2018 for Al Gore's climate leadership training, the Climate Reality Project, and then recorded the video later that year. In the poem she speaks these words slowly, so her listeners cannot miss the message: "Climate change is the single greatest challenge of our time." She continues:

So I tell you this not to scare you, /

But to prepare you, to dare you. /To dream a different reality."

It's become an accurate cliché to say that Amanda stole the presidential inaugural show two years later with the choreography of her hands, the assured rhythm of her words, her passionate grace in front of a fractured country. This charismatic young poet read her poem "The Hill We Climb," which she performed in a canary-yellow coat and wide red headband at Vice President Kamala Harris and President Joe Biden's inauguration. In later interviews she noted her favorite line from the poem: "that we've learned that the norms and notions of what 'just is' isn't always justice." Within her poetry is a revelation of how words made accessible can bring humanity to complex issues, like the climate crisis or political divisions.

> "So I tell you this not to scare you, / But to prepare you, to dare you./To dream a different reality."

As a native of Los Angeles, Amanda discovered poetry after her third-grade teacher read Ray Bradbury's "Dandelion Wine" to the class. One of three siblings, including a twin sister, she credits her mother, an English teacher, for making creativity a priority by turning off the TV and encouraging writing, playing, and fort-building. In middle school she attended a diverse private school while her mom taught public middle school, a dichotomy that made her aware of the differences dependent on zip codes and access.

As a child Amanda was challenged with a speech impediment where "Rs" were the most difficult of the sounds to formulate. "I don't look at my disability as a weakness," she told the *Los Angeles Times*. "It's made me the performer that I am and the storyteller I strive to be."

It was the nonprofit "Write Girl" that influenced her as a young writer. At fourteen years old, after reading about the

Pakistani activist Malala Yousafzai, Amanda was inspired to become a youth delegate to the United Nations.

Later as a student at Harvard, she became the first national Youth Poet Laureate, and now publishes books of her poetry for adults and children, such as *Change Sings*. Alongside speaking engagements across the country, she signed her first modeling contract to inspire fashion for a cause as a Black model. Wearing a white Prada gown, she'd again call for climate action through her poem "An Ode We Owe" at the 2022 United Nations General Assembly.

The week of the inauguration was also the first day of classes at the college where I teach. I opened the new term showing the video of "Earthrise" to my environmental education students. That night, I got an email from a student, a soccer player who'd gotten some bad news that day about an upcoming surgery.

"Tonight I went on a walk and I kept thinking about the poet you showed us in class," he said. "I've never written anything like this before. I don't know why, but I thought you would like it." He'd attached an audio recording of his first poem—about the power of the night sky to ground him. In a fractured school year and political landscape, Amanda's poetry impacted not only the country but also individual young people hoping to be brave enough to see the light and be it.

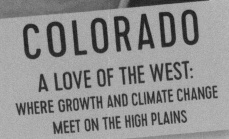

BETH CONOVER

POLICY ANALYST & NONPROFIT DIRECTOR

Denver, Colorado

What happens when a medium-sized western city like Denver increases in size by 25 percent within a decade, homes are built farther and farther up into the mountains, air pollution increases exponentially, and many Denverites aren't sure how to live sustainably in a changing climate? Beth Conover knows an answer to that. The author of *How the West was Warmed: Responding to Climate Change in the Rockies*, she's spent three decades forging links between research, policy, and conservation practice to address these questions across western cities and landscapes.

It's these connections she now brings to her role as director of the Salazar Center for North American Conservation at Colorado State University, where climate is an intersectional theme of every project applying conservation research to on-the-ground decisions faced by communities. These applications are especially relevant in Denver, formerly known as a "sleepy cow town" as recently as the 1970s, now fast becoming like southern California in its sprawl and housing prices. While the Salazar Center serves Canada, the United States, and Mexico, much of Beth's career has brought policy to practice in her hometown of Denver. She's created an intergovernmental agreement between the center and the City and County of

Denver to use taxes to fund climate actions, including green spaces and an office of climate action, sustainability, and resilience.

"City governments are overwhelmed now," she said, "They have a hard time pulling in resources. The center can help them research the risks and costs of inaction on climate and navigate the right actions to take."

Researchers at Colorado State University help to assess baseline data before the city implements specific climate actions, and then measure ongoing impacts around issues like urban tree canopy, stormwater management, transportation, air quality, and more.

As a Denver native, Beth worked as the policy director in the mayor's office from 2003 to 2007 under the leadership of John Hickenlooper, who went on to become governor and then US Senator for Colorado. She helped to craft an urban sustainability program called *Greenprint Denver* within the mayor's office, and produced Denver's first climate action plan, together with an advisory board of civic, business and environmental leaders, so this work comes full circle for her.

After growing up in the city in the 1960s and 1970s, she's the only one of her four siblings who now lives in Denver.

"As a kid, I made playhouses in trees and stayed outdoors," she said. "Even in high school, I organized this program called GORP that stood for Great OutdooRs Preservation to promote recycling at school."

From the time she was old enough to work for a paycheck, she always had a job. "My parents divorced when I was young, and it seemed like work was just what I did, in greenhouses, on trails, and more."

It was a summer job with the Youth Conservation Corps when she was only fifteen that showed her the value of public lands. "We built trails, wore hard hats, and used a Pulaski," she said, remembering how she wielded that axe-like tool to dig fire breaks to contain wildfires.

After attending public schools in Denver, she traveled to the east coast for college at Brown University, where she was exposed to "non-ironic cocktail parties and a culture that felt unfamiliar." She considered returning home, but found her people in the Environmental Studies department, where she studied in the Urban Environmental Laboratory, a training ground for her interdisciplinary work.

With that systems-focused, interdisciplinary perspective, she views the challenges faced

"Now more donors see the connection between public health, public lands, private activity and climate. It's not just the climate geeks anymore."

by the West—water, land use, fire, growth—as ones that other parts of the country grapple with and can learn from each other. "We already have climate refugees, people moving to Colorado to escape climate impacts elsewhere, and then propelling growth," she said.

Beth told me that getting people to see the value of this work was much harder three decades ago. "Back then, there were only a handful of foundations you could turn to for funding," she said. "But now more donors see the connection between public health, public lands, private activity, and climate. It's not just the climate geeks anymore."

She's figured out how her suite of skills can help the region she loves. "I had to get real clear on what I had to offer and how I could share it," she said. "I can build programs. I can raise money, and I can leverage political will." Building climate resilience in Denver will prove instructive to cities throughout the West and the world.

Newspaper headline visible in photo: **Local business** — "Herbicides banned at schools"

HAWAI'I

PLATFORM WITH A PURPOSE:
YOUTH CONFRONTING HERBICIDES AND INDUSTRY FOR HEALTH

MACKENZIE FELDMAN

EXECUTIVE DIRECTOR AND ORGANIZER

Honolulu, Hawai'i

When Mackenzie Feldman left Hawai'i to play on the beach volleyball team at the University of California, Berkeley, she had no idea her platform as a student athlete would impact every public school in her home state and public universities throughout California.

"I went to volleyball practice one day, and our coach told us to leave the balls if they rolled into the grass," she told me. "They'd just sprayed herbicides around the court."

The previous fall semester, a professor had explained that the herbicide Roundup was used throughout campus despite the fact that the World Health Organization had declared its active ingredient, glyphosate, a probable carcinogen. In addition, a warming climate indicated the likely increase in weeds and insects, prompting additional pesticide use with negative impacts on soil health and biodiversity, as well as human health.

Derived from fossil fuels, these chemicals not only destroy weeds and insects, they also destroy life in the soil and decrease its ability to absorb water. With synthetic fertilizers known to contribute to algae blooms in bodies of water, this in turn may lead to dead zones, where oxygen content is too low to support marine life.

For Mackenzie, the issue felt especially urgent since she'd grown up in Hawai'i, ground zero for industrial agriculture and GM (genetically modified) seed corn testing, resulting in an abundant use of pesticides.

"I didn't know how to start a movement against the toxic herbicide use at such a large school," said Mackenzie, known as Kenzie to friends. "But then I realized that we had a lot of power and privilege as students, and as a student athlete, I had an even bigger platform."

With her teammate Bridget Gustafson, she first met with the Athletics Grounds Manager, who didn't have the human power to weed by hand. For the area around the beach volleyball courts, their twenty-person team volunteered to do the weeding. To make their campaign official, Mackenzie and Bridget founded the organization that became Herbicide-Free Cal, organized student work days, and funded trainings led by organic landscape experts. Before she graduated in 2018, her campaign, in addition to garnering national attention about herbicides, resulted in a ban of glyphosate by all ten campuses in the University of California system.

But her organizing took an unexpected turn after she attended the 2018 trial of *Johnson v. Monsanto* in San Francisco, which pitted one school groundskeeper in California, Dewayne "Lee" Johnson, against the manufacturer of Roundup. As the first plaintiff to ever win a case against Monsanto, Lee and his legal team presented medical data, attributing his diagnosis of non-Hodgkin's lymphoma to exposure to glyphosate. After his landmark win of $289 million, the defendant appealed, and the amount was decreased to $78 million. Ultimately, Bayer, which bought Monsanto, announced it would no longer sell glyphosate-based herbicides to gardeners in the United States as of 2023.

"Lee was so brave, so I wrote to thank him for standing up to Monsanto," she told me. In response, Lee asked how he could help with her work. It was Mackenzie's sister who had the idea to fundraise for Lee and his family to share his story in Hawai'i.

"In high school, I watched as activists tried to ban GM testing and get corporations like Monsanto out of the state," she said. The very activists she'd admired as a teenager helped to make the visit possible. Individuals donated their homes for the tour of four islands, and the "Protect our Keiki Coalition" organized community events.

"*Keiki* means 'child' and this tour was also about protecting children," she explained. "The Hawai'i Board of Education chair and Superintendent had said they didn't use herbicides at public schools."

Yet at a gathering to introduce Lee to education officials, parents, and activists, an agriculture teacher said she regularly sprayed Roundup around the perimeter of the campus to control weeds and taught her students how to do it too. With the media present, the Board of Education decided on the spot to ban herbicides from all public schools in Hawai'i.

> **"I believe as my Hawai'ian ancestors taught me—there is no separation between our work and our landscape, between who we are and what we do."**

"I believe as my Hawai'ian ancestors taught me—there is no separation between our work and our landscape, between who we are and what we do," she said.

Kenzie credits her mom with teaching her about caring for food systems and the planet. After graduation, she expanded her work with Herbicide-Free Campus on a national scale with seven staff and six schools. They've rebranded as Re:wild Your Campus, a project of the organization Re:wild, with a focus on climate change, biodiversity loss, and human well-being. She also works as a Food Research Fellow with the think tank Data for Progress and helps to develop food and agriculture policy memos.

"I never considered politics in my future," she said, "but AOC [Alexandria Ocasio-Cortez] made me realize you can do this if you are young. I could see myself representing Hawai'i, especially to restore the vibrant food system [we had] before colonization." And she sees her mission—elimination of herbicides in every school in the country—as a lasting tribute to her close friend Lee Johnson.

IDAHO

TELLING CLIMATE TRUTHS:
HARNESSING STORYTELLING FOR RURAL
COMMUNITIES

JENNIFER LADINO

PROFESSOR OF ENGLISH

Moscow, Idaho

Some people argue that all fiction in the future will be climate fiction, stories that shape how we see the impact and integration of climate in our lives. An English professor at the University of Idaho, Dr. Jennifer Ladino believes reading literature allows for "safe empathy," a way to confront feelings and challenges readers might not be ready to face in real life. In her home state of Idaho, Jenn, as she's known, has organized community conversations around Barbara Kingsolver's novel *Flight Behavior*, in which climate change hits home for a small Appalachian town.

"In our rural communities in Idaho, the book gives us a way to move from talking about climate in the relatively safe realm of fiction to thinking about it in our day-to-day lives," she told me. "We've seen light bulb moments when people connect the dots between increased fires and patterns relating to climate."

This climate fiction project began in the Confluence Lab, which initiates creative interdisciplinary projects and partners with rural communities to work toward more just, sustainable, equitable futures. The origin story of the Confluence Lab includes elements we might imagine in a Stephen King novel: a stuffed polar bear, a national park, and a restored ghost town. In 2018 Jenn traveled with English department colleague

Erin James to the Taft-Nicholson Center in the Greater Yellowstone Ecosystem, a remote campus in grizzly country, and they'd been forewarned to look for a polar bear.

"The whole area is so beautiful, but like so much of the West, full of tension," Jenn said. "The Koch brothers are billionaires who own a lot of land there, so there are conservative corporate interests, but also environmental education at the center. And there's this huge taxidermied polar bear in the benefactor's house."

They left the retreat committed to creating an interdisciplinary center in Idaho to elevate people's stories related to climate and social justice. After conversations with Teresa Cavazos Cohn, the geographer who'd alerted them to the polar bear, the Confluence Lab was born. The trio initiated weekly meetings with colleagues across campus, landed space in a new research building, and soon found themselves with eight funded projects underway.

The Lab's most ambitious project to date is in partnership with the University of Oregon on the Pacific Northwest Just Futures Institute, a $4.5 million initiative to build a regional network for racial and climate justice. The Lab's contribution, "Stories of Fire: A Pacific Northwest Climate Justice Atlas," brings together educators, artists, writers, and local communities to create a multimodal atlas that gathers, tracks, and maps stories and images of wildfire. The project forges connections between fire, justice, and traditionally underrepresented rural voices.

In a separate project, Lab members gather personal stories of wildfire to enhance STEM education with storytelling. Some who shared their stories come from families who've lived on ranches or farms for decades.

"In some right-leaning, rural communities, climate change is a trigger phrase," Jenn said. "But everyone cares about pollution and their health and about being good stewards. Rural Idahoans are noticing and feeling big changes, including the new normal of longer, hotter, smokier summers."

The Confluence Lab also works to bring more complexity to the term "rural" to include Latinx, Indigenous, rural left-leaning, and rural right-leaning communities. Climate change is complicated across these demographics, but in these wildfire projects, Jenn explained, "we are looking at the capacity of [wild]fire stories to get people thinking less at an individualistic level and more about the ways we're interconnected."

For Jenn, her own story took a turn when she left the suburbs of Washington, DC, as a twenty-year-old headed to Grand Teton National Park to work at the entrance station. She returned for another twelve summers as a park ranger, falling in

"We are looking at the capacity of [wild]fire stories to get people thinking less at an individualistic level and more about the ways we're interconnected."

love with the West even as she recognized its legacy of violence, she said. When she discovered the environmental humanities, she found that intersection between land and literature and her commitment to climate justice.

But it was as a mother when she had a defining epiphany: "It was 105 degrees in June during the summer of 2015, and my kids were three and seven," she said. "They started building a 'tent village' in the basement to escape the heat. They were using the language of refugees to seek a cooler place." She realized then climate would be the focus of her work. "I know it's cliché to say I had the revelation because of being a mom, but that's what happened."

Jenn continues teaching literature that imagines present and future worlds such as Octavia Butler's *Parable of the Sower* and Louise Erdrich's *Future Home of the Living God*. It's been said that good fiction often indicts reality rather than reflects it. For rural Idaho, narratives like these reveal how personal the reality of climate change can be, and how much we share in common.

MONTANA

FOR A CLEAN AND HEALTHFUL ENVIRONMENT:
A FIFTH-GENERATION MONTANAN GOES TO COURT

GRACE GIBSON-SNYDER

STUDENT

Missoula, Montana

When she was still in high school, Grace Gibson-Snyder had a revelation: she wasn't sure if she wanted to have children. It's not that she didn't love kids and the idea of being a mother. But she couldn't imagine bringing children into a world where wildfires, drought, flooding, and extreme heat would be the norm in her state, known for its mountainous landscapes.

"I started thinking about this when I was only fifteen," she said. "I don't want my children to suffer the emotional impacts of living with more extreme weather and natural disasters."

Among Gen Z and millennials, Grace isn't alone: a study of 4,400 Americans found that one in four adults without children said the climate crisis influenced their reproductive decisions. Grace, a fifth-generation Montanan, understands the implications of climate change on future generations, including her own, many of whom are not yet old enough to vote.

In the case *Held v. State of Montana*, she's one of sixteen youth plaintiffs asserting the state violated their constitutional rights to a clean and healthful environment by supporting an energy system driven by fossil fuels, known to contribute to climate change. One of the attorneys for the case, Nate Bellinger, explained that Montana is unique among the states, in that it enshrines the right to a healthful environment for everyone.

"The constitution couldn't be clearer," he said. "The state has an affirmative duty to maintain and provide a clean and healthful environment for present and future generations."

Grace thinks most Montanans don't realize their constitution includes this explicit right. But she described nature as a "pillar of life in Montana," which motivated her to become a part of this case. Within a few miles radius of her house are two rivers, several cross-country ski areas, as well as mountain biking and hiking trails. She wants to make the most of these outdoor spaces before they change, such as the retreat of the glaciers in Glacier National Park, which may disappear in her lifetime.

"I started thinking about this when I was only fifteen. I don't want my children to suffer the emotional impacts of living with more extreme weather and natural disasters."

Her uncle is a firefighter who has battled the increasingly severe wildfires every year, including one that blanketed the area around her school in smoke.

"My freshman year in high school, there were massive fires all across the West," she remembered. "The smoke settled in the Missoula Valley as it does every year now." Her soccer team practiced inside the gym to escape the smoke, but the students with asthma couldn't play since their breathing was compromised.

In high school Grace became president of a club called "Students Against Violating the Environment," or SAVE. That's where she first heard about the organization Our Children's Trust, which provides legal support for this case and others. She emailed them to learn more and decided to become a plaintiff for the Montana case, which was filed March 13, 2020.

"Each plaintiff in the case is impacted differently, but none of us are old enough to vote," she said. "That's why this case exists in the first place."

The lack of snowpack due to warming temperatures has affected one of the plaintiffs named Sariel, whose family depends on fishing for food and their livelihood as members of the Confederated Salish and Kootenai Tribe. The family of the lead plaintiff, Rikki, owns a cattle ranch, and their livelihood has been disrupted by both flooding and drought.

The case seeks for the courts to declare that the fossil fuel energy system violates their constitutional rights. Even after the state's efforts to keep the case out of court, a Montanan judge ruled the lawsuit could proceed to trial, becoming the first youth climate trial in US history, scheduled for 2023.

I asked Grace how she talks about the case with her friends and family.

"My friends definitely care about climate change, but the details of the case don't necessarily interest them," she said. "I mean, it's kind of a strange thing to casually bring up. My extended family is very protective of Montana, so they're also very supportive."

Her mother's family came to Montana from Nebraska on the Bozeman Trail on a wagon train, six generations ago. "My great-great-great-great-great-great—well, I'm not sure how many 'greats'—grandmother wrote one of the few accounts of a female on the Bozeman Trail," she said. With that rich history, Grace awaits resolution of the trial even as her plans take shape to study international environmental policy with a possible future in the foreign service or on Capitol Hill. Wherever her future takes her, this Montanan intends to use policy to build a viable future for children of every age.

NEVADA

YOU SHOULD RUN:
STEPPING UP TO THE STATE LEGISLATURE

CECELIA GONZÁLEZ

STATE LEGISLATOR AND GRADUATE STUDENT

Las Vegas, Nevada

Cecelia González remembers three words spoken by her mentor when a seat opened up in the Nevada State Assembly: "You should run."

As a biracial community organizer, Cecelia had become friends with the first Democratic Asian legislator serving in the first female-majority state legislature in the country.

Her mentor wasn't the only one with words of encouragement. Several friends also suggested she consider running for office. As a student, she'd lobbied for reproductive justice with her state legislators and volunteered with political campaigns. "Honestly, if these men could do it, I could do this job too," she told me, reflecting on that decisive moment. "So I put my name in the hat!"

Cecelia's background reflects the constituency of District 16, where 50 percent of the population in this central Las Vegas valley includes people of color. Her mom is an immigrant from Thailand, and her father is of Mexican American descent.

"I'd grown up in Las Vegas and had campaign experience and friends, but I'm not a high-profile lawyer with money," she said. "So I pulled in my friends to do my website, graphics, and field organizing and raised about $7,000."

In the primary she ran against three white men—winning with more votes than the total number of people ever to vote in a primary before in that district. At twenty-nine she was the only woman, young person, and person of color in the race. She won in the general election in a heavily Democratic district. "You can't be progressive enough in this district, so I lucked out there," she said.

> "Honestly, if these men could do it, I could do this job too."

In the state assembly, she has a seat on the Natural Resources committee, a role of influence for climate policy. In her first legislative session in 2021, she was the primary sponsor of the 30 by 30 resolution, AJR3, which encourages state and federal governments to protect 30 percent of the land and water by 2030 to mitigate the effects of climate change and conserve biodiversity. These guidelines were drafted by the UN 2020 Convention on Biological Diversity.

After Cecelia introduced the resolution, Christi Cabrera with the Nevada Conservation League shared that the state is home to two of the fastest-warming cities in the United States, with Nevada having lost more than nine million acres of wildlife habitat due to wildfires in the past twenty years. Some lawmakers opposing the measure questioned the meaning of "protecting" land when the federal government manages more than 80 percent of land owned by the state.

"This can be a partisan issue, but communities of color are more impacted by climate change," Cecelia told me, "This resolution opens up the conversation, even if it's only a recommendation. I've learned you don't have to be an expert on an issue to do the right thing."

She traces her engagement in politics to her family and faith. Her father has been incarcerated since she was a child, and she described her visits to see him, prompting her passion for criminal justice reform. And her faith—the Afro-Cuban

Lucumi religious tradition—is one way she honors her ancestors, nature, and the land.

A personal trauma also prompted her meetings with state legislators as she advocated against sexual violence as a student. "I was sexually assaulted in college," she said, "I was in a toxic, abusive relationship, and I didn't want others on campus to have that experience." Her advocacy helped establish a 24-hour hotline for domestic and sexual abuse, and continued her graduate work in criminal justice and now her PhD in multicultural education with a focus on the school-to-prison pipeline. She ran for state office while on unemployment during the pandemic, and she was sworn into office as a biracial Latina-Asian-American woman.

"I'm the first Thai woman elected in our state," she said, "And I wanted to pay homage to my ancestors during the swearing-in ceremony." In deference to the women who'd come before her, she wore white, including a Thai traditional skirt and two white roses pinned to her lapel.

"The generation coming up is on fire for climate justice," she said. "And young people can run for office and win."

Her advice for others interested in public office is to be strategic. "I have a colleague who ran the perfect campaign as a perfect candidate, but she couldn't win in her particular district," she said. "So consider where you live and think about how you can show up in your community." Her path reminded me of the inspiring journey of one of my friends, a gay ordained minister and mother in North Carolina, who ran for a seat in the US House of Representatives against the odds.

Cecelia is also putting her support behind SB-448, an omnibus energy bill to prioritize renewable energy and union-paying jobs. "I'm a city girl who loves nature," she said. She wants to mentor other youth of color to run for public office as she serves her communities in the Nevada State Assembly and beyond.

NEW MEXICO

THE IMPORTANCE OF LIFE ITSELF:
FIGHTING FOR HEALTH ON SACRED LANDS

KENDRA PINTO

STORYTELLER AND EDUCATOR

Twin Pines, Eastern Agency of the Navajo Nation, New Mexico

It was 10:30 p.m. when Kendra Pinto got a text alerting her to a nearby explosion-turned-fire on July 11, 2016. To the west she could see an "orange glow" and heard "loud pops coming from the fire." With her father and sister, she drove to the scene to see what was happening and gauge the threat, not knowing if they should leave or stay in their home. When they arrived at the scene, Kendra soon realized these were gas storage tanks from the WPX fracking site in New Mexico. Their explosion created a fire that burned for five days.

"I then decided to stream it live on Facebook," she wrote in the *Huffington Post*. "The fire was roaring and it was roaring loud . . . All we could do was watch. Watch it be fed by oil stored at the site and by oxygen."

Families who lived closer to the site had to evacuate their Navajo sacred lands where the infrastructure of gas and oil dotted the landscape. The closest home sat less than 350 feet from the site. Indeed, the Four Corners area—where New Mexico, Arizona, Colorado, and Utah meet—has the largest concentration of methane, a byproduct of hydraulic fracking, in the United States.

Because fracking involves drilling into the earth and injecting a high-pressure mixture of water, sand, and chemicals into the rock to release natural gas, it can cause geological shifts in the ground. It requires a large amount of freshwater, and the release of methane from the process is one driver of climate change.

As a child Kendra roamed the valleys and mountains near the Chaco Canyon, an area often called "The American Cradle of Civilization," where the Anasazi lived between 900 CE and 1300 CE. She grew up in Twin Pines on the Eastern Agency of the Navajo Nation in northern New Mexico.

After living in Chicago as a young adult for several years, she returned home to care for her grandmother and discovered a transformed landscape—fracking crews, petroleum trucks, and well sites—on public lands managed by the US Department of the Interior's Bureau of Land Management. In 2020 this area alone called the San Juan Basin had between 37,000 and 41,000 active oil and gas well sites, she said.

With long brown hair and sunglasses, Kendra towers over her mother, who says proudly her daughter has inspired people around the world in her work with Diné CARE, which stands for Diné Citizens Against Ruining our Environment. Diné means "the people" in the Navajo language. During an interview for the film "The World We Want," her mom laughed: "I'm agreeing with my daughter! I'm agreeing a lot with her because she's right on a lot of things, and I am proud of her."

"We need people to care," Kendra said. After returning home, she started school at Fort Lewis College in Durango, Colorado, sleeping in her truck to avoid the three-hour round trip from home to school. She eventually had to make the commute each day to care for her grandmother at night, but even with these challenges, she earned a BA in Environmental Studies.

"When I tell people I live in the middle of nowhere, they don't really get it," she said.

To support the protection of sacred land, she began collecting air quality samples to monitor violations of leaking emissions, such as Volatile Organic Compounds, methane, and hydrogen sulfide, and testified before Congress in support of a federal law to regulate methane emissions from oil and gas drilling in 2017. During her remarks, her voice at times shook as she described her love of sacred lands and the Navajo people.

"From particular peaks, I can spot Colorado, Utah, and Arizona, all in one quick sweep," she said. "My grandma was born less than a half a mile from where she currently resides. She is ninety-two years old." She described the pollution, truck traffic, and health problems associated with the well sites, and the results from air monitoring revealing elevated hydrogen sulfide levels, whose long-term exposure is linked to respiratory health complications.

"There is nothing wrong with demanding clean air and clean water."

"There is nothing wrong with demanding clean air and clean water," she said to the elected officials. In 2019 in response to a lawsuit, a federal appeals court ruled that the Bureau of Land Management had illegally approved oil and gas drilling and fracking in the Greater Chaco region.

Since then, Kendra joined Earthworks as the Four Corners Indigenous Community Field Advocate, focusing on tribal communities adversely impacted by fracking and extraction. Reporting suspected leaks and venting emissions to the New Mexico Environmental Department is one way she highlights the inequities and violations that would go unnoticed. She also worked on a report of the mental and physical health of those in the Tri-Chapter region affected by extraction.

Sacred lands free from fracking would not only help to slow climate change, but it would also help to assure the health of those who are Diné, including Kendra and three generations of her family.

OREGON

GROWING UP IN COURT:
YOUTH SUE THE US GOVERNMENT
FOR THEIR FUTURE

KELSEY JULIANA

CLIMATE ACTIVIST

Eugene, Oregon

With a wide, beaming smile, Kelsey Juliana stood with twenty of her closest peers, her right fist raised in the air, cheeks glowing in the crisp Pacific Northwest air in front of a cheering crowd. The scene could have easily been a football game or a concert, but instead it was the steps of the Oregon federal courthouse. Youth from ten different states were suing the US government, demanding a plan to ensure their generation's constitutional right to life, liberty, and property, threatened by the escalating climate crisis. Their supporters had come to advocate for their case.

While she lives in cozy sweaters and leggings, Kelsey can just as easily don a black jacket and pants to hear testimony in court or deliver a scathing knock-down of how the federal government has known about the harmful impacts of climate change for more than fifty years—and failed to act. Her generation will be left to deal with the compromised health of both people and the planet.

Standing on the court steps to cheers of support, she was one of twenty-one coplaintiffs in the case *Juliana v. US*, informally referred to as *Youth v. Gov*, a slogan that's now representative of a global movement of youth taking litigative action to address climate. At rallies the brightly colored signs tell the story: #seas are rising, so are we; #a lawsuit can change the world; #we demand a climate recovery plan.

"This case is everything," Kelsey said on *60 Minutes*. "We have everything to lose if we don't act on climate change right now."

When Kelsey was a newborn, her parents gave her the full name "Kelsey Cascadia Rose Juliana," and took her to protests against the destruction of old growth forests, and to protect endangered species from the sale of arson-burned "salvage" timber by the federal government. By the fourth grade, Kelsey organized students at her school to participate in the first International Day of Climate Activism.

"I was blessed to have mentors and parents who were honest with me about the wellness of children and the planet," she once said. "It can be intimidating to be confronted with the reality. It's overwhelming and scary."

What's at stake is the lives of her friends and the places they call home. She's watched forest fires so intense her family had to stay inside for weeks due to the smoke surrounding the Cascade Mountains, for which she was named. Her coplaintiff Levi Draheim lives on a barrier island in Florida that could be underwater by the time he's an adult. Each youth involved in the case has a personal story of lives upended by the climate: farms lost, homes flooded, anxiety peaked.

Among this diverse group, Kelsey seems like the older sister. She's the first to laugh out loud, show raw emotion, and remind them of the long view. I saw these same qualities when she studied environmental education for a year at the college where I teach. In our community-based projects with local schools, she could dig into the research but also find joy in the journey, such as dressing up in a green tutu as the "kale fairy" for a lesson on cooking healthy food with first graders.

She knows a thing or two about staying present during uncertainty but persisting with a long-term perspective. She sued the state of Oregon, demanding climate action when she was only fifteen years old. Four years later she joined the federal lawsuit led by attorney Julia Olson, founder of the nonprofit Our Children's Trust, which represents these young people, from upstate New York to a subsistence village in Alaska.

The timeline of the case reads like a spy novel, with twists and turns to the litigation that played out while Kelsey was in high school and college. Twice, the US Supreme Court ruled the youth had the right to a trial, despite efforts by the government to dismiss the case. In 2020 a divided

> "This work must be done out of love. You cannot burn out of love."

Ninth Circuit Court found the government *had* violated their constitutional rights, but the legislative or executive branches should be responsible, not the courts. The attorneys asked for a new panel of judges given the sharp differences in their opinions. Kelsey is still waiting and fighting—using her platform in both the courts and the classroom.

Becoming a teacher is next on her horizon as education for the Earth has been a part of her life since she was young. While studying at the University of Oregon, Kelsey returned to the charter school she attended as a child. There she was asked by a student how she avoids burnout, knowing the government lied about the devastation caused by climate change. "This work must be done out of love," she said. "You cannot burn out of love."

UTAH

STRENGTH AND VULNERABILITY:
HOW WE AS WOMEN USE OUR VOICES

TERRY TEMPEST WILLIAMS

WRITER, NATURALIST, TEACHER

Castle Valley, Utah

She hadn't planned to buy leasing rights to drill for oil and natural gas on 1,120 acres of federal lands in her beloved homeland in Utah. Terry Tempest Williams was there with her husband Brooke for another reason: to protest the leasing of public lands to oil and gas companies who extract fossil fuels from the ground. But at this auction run by the Bureau of Land Management in 2016, she found herself in line with the bidders, instead of the protestors. She stayed in the line, signed a registration form, and took a number—19—and sat in the front row.

"Are you aware that if you have misrepresented yourself as a legitimate bidder with an energy company, you will be prosecuted and you could go to prison?" an agent with the Bureau of Land Management asked.

As an American citizen, she had a right to be there, she responded.

As the bidding began, the protestors in the back, which included Brooke, began singing: "People got to rise like water." At the end of the auction, after the protestors had been forced to leave, the bidders could purchase leasing rights that hadn't been sold—at a steep discount. So Terry and her husband paid

$1.50 per acre—a total of $1,168 plus a $820 processing fee—to keep fossil fuels in the ground on land near their own home.

"We put it on our credit card," she wrote in the *New York Times*. As required by federal law, they formed an energy company, Tempest Exploration Company, LLC, to draw attention to the use of public lands to fuel the climate crisis.

As a fifth-generation Mormon with deep ties to the red rock desert of southern Utah, Terry has spent her life using words and actions to protect the land and its people. This bestselling author of twenty books described finding her voice for the first time when she crossed the line at the Nevada Test Site in 1988, one year after her mother's death at age fifty-four and a year before her grandmother would die. Nine women in her family had mastectomies. Seven have died, and all had been exposed to above-ground nuclear testing.

She both celebrates and grieves her "clan of one-breasted women" in the book *Refuge*. Growing up in sight of the Great Salt Lake, she became the matriarch of her family when she was thirty. Her writing draws on the connections between our bodies and the natural world around us: "The birds and I share a natural history," she writes. My own copy of *Refuge* is filled with faded post-it notes: I first read the book as a young environmental educator and often return to her words: "As women, we hold the moon in our bellies."

As a writer and advocate for the land, she wrestles with contradictions in a measured tone: How do we protect the land if we're using fossil fuels to travel and bear witness? How can we transform our anger into "sacred rage," which can deepen into love? She takes these questions into the fight against tar sands mining in the Book Cliffs of Utah, as well as into classrooms and bookstores.

A decade ago I went to my local bookstore, where she talked about her book *When Women Were Birds*, which explores the question of why her mother left three shelves of empty journals after her death. My own mother had also died in her fifties. I'd heard Terry speak more than a decade earlier, when her hair was jet black. Now she was a luminous grey. As she stood, regal behind the podium that night, her eyes were downcast for a moment before she looked straight at the audience.

"I'm so nervous," she said.

How could *the* Terry Tempest Williams be anxious before this small-town crowd? She smiled and drew us into her fold with the intimacy of her voice. In those precious moments of introduction, she shared her challenge with stuttering as a child and asked for our patience.

> **"Once upon a time, when women were birds, there was a simple understanding that to sing at dawn and to sing at dusk was to heal the world through joy."**

She created room in our hearts for ourselves and each other for what she called "the open space of democracy," something she has pursued and protected. Then she told a story: "Once upon a time, when women were birds, there was a simple understanding that to sing at dawn and to sing at dusk was to heal the world through joy."

Eight years later she wrote me during a brutal summer of fire, smoke, drought, and the evaporation of the Great Salt Lake, a "horizon of salt" as far as she could see. The birds were few, she said, and it broke her heart. "What we thought was a pause is now a place—may it prove to be a transformative one."

WASHINGTON

CLIMATE JUSTICE FOR ALL:
YOUTH TO POWER

JAMIE MARGOLIN

ACTIVIST, AUTHOR, DIRECTOR

Seattle, Washington

We're told to listen to our elders, but young people like Jamie Margolin have placed their wisdom about the climate crisis— front and center—on the intergenerational table. They are unapologetically themselves, using their power because their lives are at stake.

"It's in my blood, caring for the living things around me," said Jamie, cofounder of Zero Hour, a youth climate justice group whose credo is giving voice to those not always heard in the fight against climate change. As a Jewish, Columbian American, queer activist from Seattle, Washington, Jamie has forged a life out of creating what didn't exist but could.

With her glittery eyeshadow, hoop earrings, brown hair streaked magenta, and a no-nonsense demeanor, Jamie came to the table with the older generation—on the split screen of Zoom—in an interview with octogenarian Jane Fonda. For this virtual conversation, Jamie sat on her childhood bed, posters behind her with messages like "We the Resilient." The actress, a red scarf around her neck, posed questions from her living room, learning from someone more than sixty years younger than herself.

"I remember wanting to take action back in second grade, but not knowing how," Jamie said. "I ended up not officially becoming an organizer until I was fourteen." At that age, she began volunteering locally with the Democratic campaign headquarters, but soon shifted to organizing around the climate.

"Climate justice has been in my bones because of where I come from in the Pacific Northwest," Jamie said. "There's just so much beauty that's being destroyed." From her mother's family in Columbia, especially her abuela, she grew up seeing their connection to the land and hearing about fish washed up in the rivers because of fracking.

Her activism was also a way to channel her clinical diagnoses of OCD, anxiety, and depression, which became especially acute after Trump's 2016 election: "So I wallowed for about a month, and then I got to work," she said. Inspired by the documentary "Awake," she saw role models in the Indigenous young women at Standing Rock protesting the Dakota Access Pipeline.

"I was like, okay, if they can do it, if they can put their bodies on the line, why am I standing here doing nothing?" So at age fifteen, she cofounded Zero Hour. Jamie is one of the torchbearers of the youth climate movement, one whose echoes can be found in women who've protested against toxic waste sites and advocated for clean water and air for children.

Frustrated by inaction among elected officials, Zero Hour's first action to center youth voices focused on organizing the 2018 youth climate march and lobby day in Washington, DC. They've continued by holding climate summits, supporting voter registration, and educating communities about the root causes of climate change, which include colonialism, capitalism, sexism, and racism.

"The carbon in the air is not racist," Jamie once said. "The people who decide where the carbon goes are racist."

She wants young people to know they don't have to attend protests to affect change. Find your own skills and apply them in your community, she writes in her book *Youth to Power: Your Voice and How to Use It*—a text I use in my own classes.

Like so many, she also experiences climate anxiety, especially after the wildfires out West. "It just hits me like a bucket of cold water, like it shocks me, and then suddenly I can't do anything," she once said. " . . . and sometimes I just have to take a little bit of a break." She calls it "escapism in small doses," but one antidote to despair for her is action.

She's a plaintiff in a lawsuit in collaboration with Our Children's Trust against the state of Washington demanding a climate recovery plan. Now a film student in college,

"I should stop wishing others will tell stories I want to be told."

she's using every tool at her disposal to confront injustice in this world, including producing TV shows and movies with queer representation. Rather than hoping Disney would create queer characters, she realized, "I should stop wishing others will tell stories I want to be told."

The person who testified before Congress is the same person who'd love to create an animated-blockbuster-queer-princess movie for Disney. "If there were no issues in the world that I had to fight against, I would be in the studio all day creating fantastical animated stories," she told me.

For her, it isn't a choice between two paths. The fight for climate justice and LGBTQ+ rights are intertwined, given the disproportionate impact of the climate crisis on the LGBTQ+ community worldwide. Her films amplify the possibility of standing up to power to craft what could be. The story of her life is to imagine what is lacking—and take concrete steps to create a more just world.

WYOMING

FROM THE ARCTIC TO THE CAMPAIGN TRAIL:
A SCIENTIST RUNS FOR THE US SENATE

MERAV
BEN-DAVID

WILDLIFE ECOLOGIST

Laramie, Wyoming

"Navigating hostile terrain has been my day job," said Dr. Merav Ben-David after she decided to run for US Senate. As a wildlife ecologist, she'd researched the impact of climate change on polar bears for twenty years and seen the Arctic ice melting over decades. At the same time, she watched politicians make decisions affecting the climate based on donor demands rather than science. In short, she took up the challenge.

For starters, all campaign events were virtual in 2020, and she didn't have a Facebook or Instagram account before her candidacy. She was running for a national office without prior experience as an elected official. And perhaps the largest of all: she lived in the sparsely populated state of Wyoming, which hadn't elected a Democrat to federal office since 1976.

But Merav has held a sedated polar bear in her arms in the Arctic to gather physiological data and studied the impact of the Exxon Valdez oil spill on wildlife in the Gulf of Alaska.

"I realized I can't leave it for others to do," she told me. "Those of us who know what is happening with the climate must run for office and make decisions with the science in mind. It is now or never."

While she won the Democratic primary by 40 percent of the vote, she was the underdog against Republican Cynthia

Loomis, a former state legislator and representative to the US Congress. Yet she hasn't shied away from a high-stakes gamble, at the University of Wyoming, where she teaches, or in her hometown of Rishon Le-Zion, Israel.

"I was born and raised on a family farm," she told me. "Everyone in the village knew that if they found an animal who was seriously injured, they could bring it to me as I'd nurse it back to health and release it."

> **"Those of us who know what is happening with the climate must run for office and make decisions with the science in mind. It is now or never."**

Birds, a baby hedgehog, a gazelle fawn, a small porcupine: these animals found themselves in the hands of a young wildlife rehabilitator dedicated to their lives, against all odds.

She served two years in the military, studied biology and zoology in Tel Aviv for her undergraduate and master's degrees, and then received funding to pursue her PhD in wildlife management at the University of Alaska, Fairbanks, where she researched the effects of salmon on mink and marten. It was her dissertation advisor who supported her application for permanent residency and ultimately citizenship in the United States. During her post-doc, she studied the impacts of the Exxon Valdez oil spill on wildlife.

In her thirty-year career, Merav had a front-row seat to the warming waters and the melting of the sea ice. On her Twitter feed, I read a quote from Polar Bears International, which I couldn't forget as we chatted. "Sea ice is to the Arctic as soil is to the forest."

"One of the things we ecologists need to emphasize is the value of long-term research," she said. "Ecological systems are chaotic in nature, and small changes can lead to variation. But when you are there for twenty or thirty years, then the trends show up."

"We are there as scientists," she said, emphasizing every word. "We see with our own eyes."

When she talked with elders in Indigenous coastal communities in Alaska, they told her they'd never seen rain before in October. "So it's not just me with my satellite records."

She summarized the situation for me. "Polar bears need ice. The ice is disappearing faster than we ever predicted. The Arctic is heating up faster than anyplace in the world. We are melting the polar bear's habitat."

Polar bears are adapted to go without food for up to 180 days, but there is a threshold beyond which they can't survive.

"The question is: Do we even care?" she said. "It's about our future as well as the future of wildlife."

Although she'd published more than one hundred scientific articles, Merav wanted to make a difference in Wyoming—a state receiving more than half its revenue from the fossil fuel industry, and one that has struggled economically. So she campaigned on a platform to "reimagine and rebuild," emphasizing strategies to strengthen the economy while addressing the climate crisis.

While her opponent won with 60 percent of the vote, Merav considered the experience with the long-term view of the field researcher she's been for most of her life. "I learned so much about how a campaign is run," she said. "And now I'm supporting other candidates who are planning campaigns to promote climate solutions."

Would she run again? Yes, but it would depend on the race, she said. The biggest achievement of her campaign was that she helped to change the conversation at a state level. "People weren't talking about climate change, and now they are," she said.

LOOKING FORWARD
THIS LOVE STORY ISN'T OVER YET

There are many points of connection revealed in these fifty climate stories from across the land. Each woman reminds me again and again that I am not alone, and this is not the end of the story. No way. No how. I need to hear this truth to shore up courage for the long haul, especially since my ordinary days are consumed with the logistics of life as a mother, sister, teacher, and friend—in a climate emergency.

So take a deep breath and repeat with me: We are not alone. And it's not over yet.

In every single story, one theme stands out to me: these women are damned good at working together. They know collective momentum is the only way to disrupt the power structures invested in the status quo of our patriarchal, capitalist fossil fuel economy. Their stories show me how we ground and sustain ourselves both in daily life and in a crisis. With each other, amplifying power, we can help to imagine and reshape a more just world in the midst of uncertainty. As Rebecca Solnit says, "Imagination is a superpower."

But this is not a solo job. From these women, I've seen how we stand on the shoulders of those who've come before us, but also how we can link arms with those who cultivate our strengths and steady us to do the next best thing.

No one in this book is a picture-perfect climate hero (Thank goodness! Who wants those kinds of friends?) These women lead messy, complicated lives, like us all, and they aren't afraid to bare real emotions—at the kitchen table, in the courtroom, by the riverside. Inspired by others, they seem to be leading imperfect but authentic lives rooted in love. And after hearing these stories, I started to see that same legacy of love in my past.

My mother, who died at the age of fifty-eight in a tragic biking accident, was the least judgmental person I've ever met. But as a Southern woman, she had a simple—and disarmingly effective—way of letting me know if she disapproved of someone I was hanging out with as a teen.

"You know, Mallory," she'd say in an off-handed way, as if sharing family gossip or adding an item to a grocery list. "That person doesn't bring out the best in you."

My mother, whom I miss with all my heart, was usually right. These days, as a parent of a twenty-three and sixteen-year-old, I yearn for her guidance, which I can only find in her stories. Right now more than ever, it matters who we link arms with. We need those who listen to facts and aren't afraid to act. We need our best selves in collective to hold each other up in our intersectional world where everything is connected, and our basic rights are threatened. Silos are sabotage in a climate emergency.

Climate journalist Amy Westervelt described this view of the world after her father killed himself in the midst of the pandemic: "Maybe these things sound completely unrelated—climate change, mental health care, childcare—but to me, they are all layers to the same uniquely American issue: the idea that each of us can and should solve systemic problems on our own."

She nailed my individual exhaustion with her words: "The cult of personal responsibility is killing us."

In this climate movement, we have the power to bring people into life-giving collaboration and joy, rather than drive individuals away with fatalism and doom, according to Drs.

Ayana Elizabeth Johnson and Katharine Wilkinson. Their photo on the book jacket of *All We Can Save*, which I cite often, shows them skipping together and leaping into the air. Such joy is not borne of naivete but of a raw strategy for living in our bodies. It's a way of life that sees rest and relaxation as a balm for anxiety and activation over time. I want these people on my team. We've got to hang with those who build us up for the long haul, just like my mom advised.

As Leah Thomas, author of *The Intersectional Environmentalist*, writes, "Call it rose-colored goggles, but I find peace in imagining total liberation of both my people and the planet, and that's why I identify as a Black climate optimist and futurist."

So how do we imagine the future since daily to-do lists can be staggering, even without a climate crisis? How do we connect with women in this book or those who live in our apartment buildings, vote at our polling places, or show up in our Instagram reels? How do we join with others in a way that gives us power, rather than drains us? In *Time Magazine*, Katharine Wilkinson lays out five steps—a call to climate action and reflection rooted in community.

1. Feel your feelings: What do you feel is at stake?
2. Scout your superpowers: What knowledge, skills, and resources could you contribute?
3. Survey solutions: What climate solutions capture your imagination?
4. Consider your context: What opportunities exist for you to act in your personal, professional, and public life?
5. Cultivate a climate squad: Where are people you can link arms with?

Your skills may be as an artist, an activist, an engineer, a politician, a friend, a mother, an aunt, a poet, a weaver, a singer, or a cook. We need you with us, your climate squad.

And if you're not sure what to do, recall the words of Dr. Katharine Hayhoe: The best thing you can do about climate is to talk about it. Tell your story—not to the 10 percent of the US population dismissive of climate change but to the other 90 percent. Listen to the stories of others. Bond over shared values and connect climate to what you both care about. Climate storytelling reveals the heartache and joy of living right now. It allows us to see our humanity in each other and imagine a world we want—while we learn how to take care of ourselves and each other, even when we don't know how the story will end.

At the college where I teach, three words are sewn on a banner—*Love Above All*—words to live by for this Earth and each other. I think my mother embodied these words, as do the women who shared their stories with me. And with help, I can too. The bottom line is this: it's not too late, and our love story is not over yet.

GRATITUDE

Many good ideas begin around kitchen tables. This book is one of them. To Jill Drzewiecki, whose clear vision and steadfast love I'll follow forever. I'm grateful for your unrelenting work for girls and women, from refugee camps in Malawi to your base in Wisconsin and your other home in North Carolina. Thank you for walking into my basement office more than twenty years ago and inviting me to sit at your table.

To each of the women who talked with me, reviewed drafts, shared photographs, and allowed me to connect with your story. You made me feel like a member of the climate-solutions paparazzi: I gained friends and momentum from your presence. As the world faced a pandemic, hurricanes, wildfires, war, loss of reproductive rights, and other devastations, you highlighted one truth for me: finding joy in the journey builds collective stamina for the long haul.

To my students at Warren Wilson College. To LeeAnne Beres, Liesl Erb, and Dayna Reggero, who took time to share names and lift up many of the stories in this book. To Katharine Wilkinson, for allowing me to share her five steps for climate action. To Brian Cole, who models friendship grounded in deep faith and light laughter. To Lyn O'Hare, who teaches me that curiosity about others might save us from ourselves. To my sister Margaret, who defines family to me. To my mother Ann, whose vibrant life and untimely death remind me of what matters. To my daughters, Maya and Annie Sky, my circle of love in a big, wide world.

GRATITUDE

To my writing mentor Jill Rothenberg. To Carol Mann, my agent. To Lil Copan, my editor, and the team at Broadleaf Books, for their mission to publish for the common good in community.

To the Tsalagi/Cherokee, on whose land I live in these valleys and mountains of Western North Carolina. And lastly, to the Earth and Mother who sustains us all. May love and light perpetual shine upon Her.

SELECT RESOURCES

In addition to the resources below, the bibliography includes all secondary sources used as background research for each story.

Good Energy. *Good Energy: A Playbook for Screenwriting in the Age of Climate Change*. 2022. https://www.goodenergystories.com/.

Gunn-Wright, Rhiana. "A Green New Deal for All of Us." In *All We Can Save: Truth, Courage, and Solutions for the Climate Crisis*, edited by Ayana Elizabeth Johnson and Katharine Wilkinson (New York: One World, 2020), 92–201.

Hayhoe, Katharine. *Saving Us: A Climate Scientist's Case for Hope and Healing in a Divided World*. (New York: Atria/One Signal, 2021).

Houska, Tara. "The Standing Rock Resistance and Our Fight for Indigenous Rights." TED Talk. 2017. https://www.ted.com/talks/tara_houska_the_standing_rock_resistance_and_our_fight_for_indigenous_rights?language=en.

Johnson, Ayana Elizabeth. "How to Find Joy in Climate Action". TED Talk. 2022. https://www.ted.com/talks/ayana_elizabeth_johnson_how_to_find_joy_in_climate_action.

SELECT RESOURCES

Johnson, Ayana Elizabeth, and Katharine Wilkinson, eds. *All We Can Save: Truth, Courage, and Solutions for the Climate Crisis.* (New York: One World, 2020).

Lockwood, Devi. *1001 Voices on Climate Change: Everyday Stories of Flood, Fire, Drought, and Displacement from Around the World.* (New York: Simon & Schuster, 2021).

Malala Fund. *A Greener, Fairer Future: Why Leaders Need to Invest in Climate and Girls' Education.* March 2021. https://malala.org/newsroom/archive/malala-fund-publishes-report-on-climate-change-and-girls-education.

Our Children's Trust. "Securing the Legal Right to a Safe Climate." 2021. https://www.ourchildrenstrust.org/.

Project Drawdown. "Drawdown: The World's Leading Resource for Climate Solutions." 2021. https://drawdown.org/.

Reggero, Dayna. *Planet Prescription: Mothers and Others for Clean Air. Climate Listening Project.* 2020. https://planetprescription.com/about.

Solnit, Rebecca. "Ten Ways to Confront the Climate Crisis without Losing Hope." *The Guardian.* Nov. 18, 2021. https://www.theguardian.com/environment/2021/nov/18/ten-ways-confront-climate-crisis-without-losing-hope-rebecca-solnit-reconstruction-after-covid.

Toney, Heather McTeer. "Black Women Are Leaders in the Climate Movement." *New York Times.* June 25, 2019. https://www.nytimes.com/2019/07/25/opinion/black-women-leaders-climate-movement.html.

PARTNERSHIPS
AND ORGANIZATIONS

Taking action for climate justice is more effective with others. The stories in this book reference climate-focused organizations that readers can engage with both in person and online. Getting involved can start with checking out a website, attending a meeting, or just asking about what's happening in your community.

350.org and 350 Maine

All We Can Save

Appalachian Voices

Association of Nature and Forest Therapy

Chisholm Legacy Project

CLEO Institute—Climate Leadership (through) Engagement Opportunities

Climate Cardinals

Climate Center at Texas Tech University

Climate Imperative

Climate Listening Project

Climate Reality Project

Community Farm Alliance

Confluence Lab at the University of Idaho

Creation Care Alliance

Diné CARE—Citizens Against Ruining our Environment

equity² LLC

Earthworks

Fridays for the Future

Giniw Collective

Good Energy

PARTNERSHIPS AND ORGANIZATIONS

Green New Deal Network

GreenFaith

Greenpeace

Gulf Coast Center for Law and Policy

Gwich'in Steering Committee

Heifer USA

Herbicide-Free Campus

Indigenous Environmental Network

Institute for the Study of Earth, Ocean, and Space, University of New Hampshire

Interfaith Power & Light with affiliates in 40 states

International Indigenous Youth Council

MADRE

Maine Youth for Climate Justice

Moms Clear Air Force

NAACP

Nevada Conservation League

New Consensus

Our Children's Trust

Outrider Foundation

Pacific Northwest Just Futures Institute

Project Drawdown

Roosevelt Institute

Salazar Center for North American Conservation at Colorado State University

Sunrise Movement

The Nature Conservancy

The Sierra Club

Zero Hour

50 WAYS TO LOVE YOUR MOTHER

There are thousands of strategies to unite for climate justice and love the Earth. Inspired by these women, I gleaned fifty ways—big and small—which stood out to me. Some might take a few minutes of reflection while others could take years of collective action. Distilling a one-liner from the story of a life may seem reductionist, but a list can be a launchpad for imagination and joy. Sometimes it's a good place to start. For more context, check out each story.

SOUTH

Alabama

Look for TV shows and films with realistic and accurate climate narratives (and watch them or even write them).

Arkansas

Find your own North Star—a higher purpose or mission—that helps you steward food systems, the land, and the planet.

Delaware

Join climate conversations with people of different ages, faiths, and backgrounds.

Florida

Support the engagement of youth in local environmental education programs.

Georgia

Find opportunities for collaborative and creative leadership that centers girls and women.

Kentucky

Celebrate and invest in the contributions of BIPOC healers, farmers, and business owners.

Louisiana

Advocate for cities to invest in equitable infrastructure for climate resilience.

Maryland

Recognize that advocacy for civil rights and climate justice are profoundly linked together.

Mississippi

Explore the intersection of climate with health, wellness, and equity to protect all children.

North Carolina

Listen to climate stories, ask questions, and share your story with others.

Oklahoma

Spend time outside considering the inherent rights of nature to exist and thrive.

South Carolina

Use your voice even when you're scared.

Tennessee

Research options for sustainable fibers and fashion.

Texas

Talk about climate in your everyday life by connecting around shared values.

Virginia

Pay attention to the outlets where people can access reliable climate information.

West Virginia

Advocate for renewable energy.

NORTHEAST

Connecticut

Recognize emotions and emotional responses as appropriate reactions to climate change.

Maine

Connect youth and adults together to protect climate and communities.

Massachusetts

Amplify the power of youth demanding their right to sustainable jobs and a viable future.

New Hampshire

Observe the connections between economy, recreation, and climate in your community.

New Jersey

Explore religious and spiritual values across faith traditions that align with climate justice.

New York

Protect the critical role of oceans in the climate system.

Pennsylvania

Expect clean air and water as a right for all children.

Rhode Island

Frame climate change around shared identities beyond political affiliation, such as region, recreation, vocation, family roles, and more.

Vermont

Ask someone to tell you a story about water or climate and see what unfolds.

MIDWEST

Illinois

See all fiscal policy as climate policy.

Indiana

Use time in nature to connect to the biodiversity around you and heal body and spirit.

Iowa

Go outside for twenty minutes and look up at the clouds, sit on a bench, or take a walk.

Kansas

See the sun as a source of power and light that can fuel communities, including infrastructure like transportation.

Michigan

Speak up for anti-racism and environmental justice no matter how young or old you are.

Minnesota

Respect and support Indigenous voices and sacred lands.

Missouri

Consider opportunities for impact investing in regional economies.

Nebraska

Make art for the Earth and each other.

North Dakota

Envision a world with community gardens, energy from the sun and the wind, and jobs for all.

Ohio

Hold elected leaders accountable for a future where renewable energy replaces the fossil fuel economy.

South Dakota

Do the next right thing to help protect Indigenous values and land.

Wisconsin

Study the legacies of diverse environmental leaders who have come before us.

WEST

Alaska

Pressure banks and other institutions to protect the climate and divest from fossil fuels.

Arizona

Value scientific facts and climate research.

California

Read poetry and write poems to help us see a wider world.

Colorado

Recognize that cities can share lessons from innovations in infrastructure, land use, food security, biodiversity conservation, and transportation.

Hawai'i

Tap into your unique power and platform to affect climate policies and practices around you.

Idaho

Explore literature that integrates climate reality, from fiction to YA to graphic novels.

Montana

Follow and support court cases of youth suing the government for the constitutional right to a healthy life.

Nevada

Encourage diverse voices to run for public office at the local, state, and federal levels.

New Mexico

Form partnerships to quantify the impact of fossil fuel industries with monitoring data as a tool.

Oregon

Leverage family history and legacies in one place to conserve forests, rivers, and air in that region.

Utah

Advocate for the protection of public lands.

Washington

Create what you want to see if it doesn't exist.

Wyoming

Value your experience and expertise to affect political conversations, from the local to the national level.

> **.....And finally, remember to BREATHE and REST**. As Resmaa Menaskem says in *My Grandmother's Hands*, "No human body can be activated all the time ... Help it settle, over and over. Have a bit of fun, now and then."

BIBLIOGRAPHY

The references below include secondary sources used as background research for each climate story, in addition to primary sources such as personal interviews. The states are listed here in alphabetical order.

Introduction: Love Above All

Bastida, Xiye. "Calling in." In *All We Can Save: Truth, Courage, and Solutions for the Climate Crisis*, edited by Ayana Elizabeth Johnson and Katharine Wilkinson (New York: One World, 2020), 3–7.

Carrington, Damian. "Christiana Figueres on the Climate Emergency: 'This is the decade and we are the generation.'" *The Guardian*. February 15, 2020. https://www.theguardian.com/environment/2020/feb/15/christiana-figueres-climate-emergency-this-is-the-decade-the-future-we-choose?fbclid=IwAR2ZkE19rhz5dpK0TsUdMqye6aqA74qGCd6iw-yaR6pP0C-I1fSEDGHVO7c

Dalton, Greg. "Inconspicuous Consumption: The Environmental Impact You Don't Know You Have." *Climate One*. Commonwealth Club. January 10, 2020. https://www.climateone.org/audio/inconspicuous-consumption-environmental-impact-you-dont-know-you-have.

BIBLIOGRAPHY

Heglar, Mary Annaïse. "But the Greatest of These Is Love." *Medium*. July 17, 2019. https://medium.com/@maryheglar /but-the-greatest-of-these-is-love-4b7aad06e18c.

Johnson, Ayana Elizabeth, and Katharine Wilkinson. "Begin." In *All We Can Save: Truth, Courage, and Solutions for the Climate Crisis*, (New York: One World, 2020), xvii-xxiv.

Malala Fund. *A Greener, Fairer Future: Why Leaders Need to Invest in Climate and Girls' Education*. March 2021. https://malala. org/newsroom/archive/malala-fund-publishes-report-on-climate-change-and-girls-education.

"Rhiana Gunn-Wright on Fighting for Climate Justice." Interview with Bloomberg Live. July 22, 2020. https://www. youtube.com/watch?v=ZctytHVY52s.

Solnit, Rebecca. "When the Hero Is the Problem: On Robert Mueller, Greta Thunberg, and Finding Strength in Numbers." *LitHub*. April 2, 2019. https://lithub.com/ rebecca-solnit-when-the-hero-is-the-problem/.

Wright, Georgia. "The News Is Bad—So Take Care of Yourselves." *Hot Take Newsletter*. August 8, 2021.

ALABAMA—Anna Jane Joyner

Eyen, Lena. "Anna Jane Joyner: When the Apple Falls Far from the Tree." Markkula Center for Applied Ethics. Santa Clara University. April 1, 2017. https://www.scu.edu/ environmental-ethics/environmental-activists-heroes-and-martyrs/anna-jane-joyner.html.

Good Energy. *Good Energy: A Playbook for Screenwriting in the Age of Climate Change*. 2022. https://www.goodenergystories. com/.

McPherson, Coco. "God's Work: Meet the Woman Turning Evangelicals into Environmentalists." *Rolling Stone*.

June 15, 2015. https://www.rollingstone.com/politics/
politics-news/gods-work-meet-the-woman-turning-
evangelicals-into-environmentalists-68581/.

Schigeoke, Scott. "Undaunted: The Climate Activist Who Hasn't
Given Up on Mainstream America." *Grist*. December 23,
2019. https://grist.org/climate/the-climate-activist-who-
hasnt-given-up-on-mainstream-america/.

The Years Project. "Preacher's Daughter—with Ian Somer-
halder." Showtime, 2014. http://web.archive.org/
web/20210802032419/https://theyearsproject.com/
story/preachers-daughter/.

ALASKA—Bernadette Demientieff

Demientieff, Bernadette. "Protecting the Arctic Refuge Is
Non-negotiable." *Patagonia*. August 17, 2017. https://
www.patagonia.com/stories/protecting-the-arctic-ref-
uge-is-non-negotiable/story-32929.html.

————. "Gwich'in Leaders Travel to New York to Tell Banks:
Defend the Arctic Refuge." *Medium*. October 29,
2018. https://medium.com/@bernadettedemientieff/
gwichin-leaders-travel-to-new-york-to-tell-banks-
defend-the-arctic-refuge-c2d09e2afdb.

————. "Testimony to the US House of Representatives Committee on Natural Resources, Subcommittee on Energy
and Mineral Resources." March 26, 2019. https://www.
congress.gov/116/meeting/house/109126/witnesses/
HHRG-116-II06-Wstate-DemientieffB-20190326.pdf.

Future Coalition. "Big Banks, the Gwich'in Nation, & the
Fight to Protect the Arctic Refuge." Earth Day Live.
YouTube. April 23, 2020. https://www.youtube.com/
watch?v=LL_J51DOQLg.

Gwich'in Steering Committee. "Protecting the Sacred Place Where Life Began." Gwich'in Steering Committee. 2021. https://ourarcticrefuge.org.

Hanlan, Tegan and Nat Herz. "Major Oil Companies Take a Pass on Controversial Lease Sales in Arctic Refuge." *NPR*. January 6, 2021. https://www.npr.org/2021/01/06/953718234/major-oil-companies-take-a-pass-on-controversial-lease-sale-in-arctic-refuge.

Our Daily Planet. "Our Hero of the Week: Bernadette Demientieff, ED of the Gwich'in Steering Committee." December 19, 2019. https://www.ourdailyplanet.com/story/our-hero-of-the-week-bernadette-demientieff-ed-of-the-gwichin-steering-committee/.

ARIZONA—Diana Liverman

Kapoor, Maya. "Geographer Diana Liverman Explains How to Tackle the Climate Crisis Fairly." *High Country News*. March 19, 2021. https://www.hcn.org/issues/53.5/ideas-interview-geographer-diana-liverman-explains-how-to-tackle-the-climate-crisis-fairly.

Liverman, Diana. "Dr. Diana Liverman Presenting at the Nobel Conference 55: How Can We Respond to Climate Change and Meet Our Goals for Sustainable Development?" Gustavus Adolphus College. YouTube. September 24, 2019. https://www.youtube.com/watch?v=PdO80W2nnrQ.

University of Arizona. "2011 Regents Professor Diana Liverman." YouTube. March 29, 2012. https://web.archive.org/web/20210723192637/https://www.youtube.com/watch?v=cdSICHymh4Q.

Yeo, Sophia. "Fixing Sexism at the Intergovernmental Panel on Climate Change." *Pacific Standard Magazine*. July 20,

2018. https://psmag.com/environment/fixing-sexism-at-the-intergovernmental-panel-on-climate-change.

ARKANSAS—Donna Kilpatrick

Bauder, Penny. "Social Impact Heroes Helping Our Planet: How and Why Donna Kilpatrick of the Heifer Ranch Decided to Change the World." *Authority Magazine.* August 13, 2020. https://medium.com/authority-magazine/social-impact-heroes-helping-our-planet-how-and-why-donna-kilpatrick-of-the-heifer-ranch-decided-b8a7b840f355.

Bergman, Annie. "Regenerative Agriculture Is Transforming Heifer Ranch into the Garden of Eden." Heifer International. September 15, 2020. https://www.heifer.org/blog/regenerative-agriculture-is-transforming-heifer-ranch-into-the-garden-of-eden.html.

CALIFORNIA—Amanda Gorman

Barajas, Julie. "How a 22-year-old L.A. Native Became Biden's Inauguration Poet." *The Los Angeles Times.* January 17, 2021. https://www.latimes.com/entertainment-arts/books/story/2021-01-17/amanda-gorman-biden-inauguration-poet.

Brown, Maressa. "Who Is Amanda Gorman? How the Young Poet Became the Star of Inauguration Day." *Parents Magazine.* January 20, 2021. https://www.parents.com/news/who-is-poet-amanda-gorman-what-you-need-to-know-about-the-national-youth-poet-laureate/.

Elan, Priya. "Amanda Gorman Signs Modeling Contract After Star Turn at Inauguration." *The Guardian.* January 27, 2021.

https://www.theguardian.com/us-news/2021/jan/27/ amanda-gorman-img-models-youth-poet-laureate.

Gorman, Amanda. "Earthrise: A Poem by Amanda Gorman." North American Association of Environmental Education. January 1, 2019. https://naaee.org/eepro/blog/ earthrise-poem-amanda-gorman.

Hertsgaard, Mark. "Amanda Gorman's Poem Rhymes with Biden's Climate Agenda." *The Nation.* January 22, 2021. https://www.thenation.com/article/environment/ amanda-gorman-climate-biden/.

Murphy, Hannah. "Amanda Gorman Is on a Mission." *Rolling Stone.* March 1, 2019. https://www.rollingstone.com/ culture/culture-features/amanda-gorman-youth-poet-laureate-800016/.

COLORADO—Beth Conover

Conover, Beth, ed. *How the West Was Warmed: Responding to Climate Change in the Rockies.* (Golden, CO: Fulcrum Publishing, 2009).

Salazar Center for North American Conservation. 2021. https:// salazarcenter.colostate.edu/.

Salit, Richard. "Continental Scale: Beth Conover '87 Heads up a New Policy Center with Far-reaching Conservation Goals." *Brown Alumni Magazine.* March 19, 2019. https://www.brownalumnimagazine.com/articles/ 2019-03-19/continental-scale.

CONNECTICUT—Wanjiku "Wawa" Gatheru

Adams, Genetta, and Janelle Dixon. "Young Futurists 2020: Wanjiku Gatheru." *The Root.* https://www.theroot.

com/young-futurists-2020-america-needs-leaders-now-more-th-1842641788#wanjiku–gatheru.

BlackGirlEnvironmentalist. Instagram page. https://www.instagram.com/blackgirlenvironmentalist/?hl=en.

Gatheru, Wanjiku. "It's Time for Environmental Studies to Own Up to Erasing Black People." *Vice.* June 11, 2020. https://www.vice.com/en/article/889qxx/its-time-for-environmental-studies-to-own-up-to-erasing-black-people.

———. "Meet Wawa." 2021. https://wawagatheru.org.

Kline, Jennifer. "Here's Why Climate Change Fear Can Feel like Grief." 2020 Makers Conference. 2020. https://www.intheknow.com/post/heres-why-climate-change-fear-can-feel-like-grief/.

Ruf, Jessica. "Why Environmental Studies Is among the Least Diverse Fields in STEM." *Diverse Issues in Higher Education.* February 16, 2020. https://www.diverseeducation.com/institutions/hbcus/article/15106248/why-environmental-studies-is-among-the-least-diverse-fields-in-stem.

Singer, Jenny, and Mattie Kahn. "Meet Glamour's 2020 College Women of the Year." July 1, 2020. https://www.glamour.com/story/meet-glamours-2020-college-women-of-the-year.

DELAWARE—Lisa Locke

Delaware Interfaith Power and Light: "Community Climate Conversations." 2021. https://delawareipl.org/dev/climate-conversations/.

Delaware Sierra Club. "Live with Lisa Locke of Delaware Interfaith Power and Light." 2020. https://www.facebook.com/watch/live/?v=862956104108326&ref=watch_permalink.

FLORIDA—Caroline Lewis

Granfield, Caitlin. "Longtime Coconut Grove Educator and Activist Dubbed 'Jane Goodall of Climate Change.'" December 1, 2016. https://account.miamiherald.com/paywall/subscriber-only?resume=118327733&intcid=ab_archive.

Key Biscayne Community Foundation. "CLEO Institute, Caroline Lewis, Citizen Science 2020." February 29, 2020. YouTube. https://www.youtube.com/watch?v=A2MrHtHnaTg.

Reggero, Dayna, and Moms Clean Air Force. *The Story We Want. Reality: The Power of Now.* Facebook Live. January 18, 2018. https://www.facebook.com/watch/?v=1676172799108473.

Time Magazine. "Meet the 31 People Who Are Changing the South." July 26, 2018. https://time.com/5349036/people-changing-the-south/.

Voyage MIA: Miami's most inspiring stories. "Meet Carolina Lewis of the CLEO Institute." January 17, 2019. http://voyagemia.com/interview/meet-caroline-lewis-cleo-institute-miami/.

GEORGIA—Katharine Wilkinson

Hawken, Paul, ed. *Drawdown: The Most Comprehensive Plan Ever Proposed to Reverse Global Warming.* (New York: Penguin, 2017).

Jackson, Lauren. "The Climate Crisis Is Worse for Women. Here's Why." *New York Times.* Aug. 24, 2021. https://www.nytimes.com/2021/08/24/us/climate-crisis-women-katharine-wilkinson.html.

BIBLIOGRAPHY

Johnson, Ayana Elizabeth, and Katharine Wilkinson, eds. *All We Can Save: Truth, Courage, and Solutions for the Climate Crisis.* (New York: One World, 2020).

Joyner, Anna Jane, and Mary Anne Hitt. "Women, Faith, and Courage for the Broken-hearted with Dr. Katharine Wilkinson." *No Place Like Home podcast.* May 6, 2020. https://podcasts.apple.com/us/podcast/women-faith-courage-for-brokenhearted-dr-katharine/id1158028749?i=1000473753473.

Wilkinson, Katharine. *Between God and Green: How Evangelicals are Cultivating a Middle Ground on Climate Change.* (Oxford: Oxford University Press, 2012.)

———. "How Empowering Women and Girls Can Help Stop Global Warming." TEDWomen 2018. https://www.ted.com/talks/katharine_wilkinson_how_empowering_women_and_girls_can_help_stop_global_warming?language=en.

———. "Women, Girls, and Non-binary Leaders Are Demonstrating the Kind of Leadership Our World So Badly Needs." *The Elders.* December 6, 2019. https://theelders.org/news/women-girls-and-non-binary-leaders-are-demonstrating-kind-leadership-our-world-so-badly-needs.

HAWAI'I—Mackenzie Feldman

Gillam, Carey. *The Monsanto Papers: Deadly Secrets, Corporate Corruption, and One Man's Search for Justice.* (Washington, DC: Island Press, 2021).

"MacKenzie Feldman, 2019 Brower Youth Awards." YouTube. October 17, 2019. https://www.youtube.com/watch?v=wNpXc0bZa3c.

"Protect Our Keiki Coalitions Brings DeWayne Lee Johnson to Hawai'i". YouTube. September 30, 2019. https://www.youtube.com/watch?v=yvjJvECuDww.

IDAHO—Jennifer Ladino

Confluence Lab. "Confluence Lab." 2022. https://www.theconfluencelab.org.

Ladino, Jennifer. "Fear in Climate Change Fiction by Jennifer Ladino." YouTube. June 25, 2020. https://www.youtube.com/watch?v=drYBSK1Lh-g.

University of Idaho. "Confluence Lab Partners Address Pacific Northwest Justice Issues with $4.5 Million Grant." January 14, 2021. https://www.uidaho.edu/news/news-articles/news-releases/2021-spring/011421-confluencelab.

ILLINOIS—Rhiana Gunn-Wright

Gunn-Wright, Rhiana. "A Green New Deal for All of Us." In *All We Can Save: Truth, Courage, and Solutions for the Climate Crisis*, edited by Ayana Elizabeth Johnson and Katharine Wilkinson (New York: One World, 2020), 92–201.

Gunn-Wright, Rhiana, Kristin Karlsson, Kitty Richards, Bracken Hendricks, and David Arkush. "A Green Recovery: The Case for Climate-forward Stimulus Policies in America's COVID-19 Recession Response." A report for the Roosevelt Institute. October 29, 2020.

"Rhiana Gunn-Wright Endorses Elizabeth Warren for President." YouTube. January 13, 2020. https://www.youtube.com/watch?v=5uMlolW6la4.

"Rhiana Gunn-Wright on Fighting for Climate Justice." Bloomberg Green Virtual Event: The Time is Now. YouTube. July 22, 2020. https://www.youtube.com/watch?v=ZctytHVY52s&t=4s.

"Rhiana Gunn-Wright Is Fighting the Policies Making People Sick." Bloomberg Green. August 13, 2020. https://www.bloomberg.com/news/videos/2020-08-13/rhiana-gunn-wright-is-fighting-the-policies-making-people-sick-video.

Owens, Donna. "She, the People: Meet Rhiana Gunn-Wright, an Architect Behind the Green New Deal." *Essence*. April 17, 2019. https://www.essence.com/feature/she-the-people-rhiana-gunn-wright-green-new-deal/.

INDIANA—Lou Weber

Einhorn, Catrin. "Our Response to Climate Change Is Missing Something Big, Scientists Say." *New York Times*. June 10, 2021. https://www.nytimes.com/2021/06/10/climate/biodiversity-collapse-climate-change.html.

Weber, Lou. "An Ecologist's Guide to Nature Activity for Healing." *Ecopsychology* 12 (September 2020): 231–235. http://doi.org/10.1089/eco.2019.0077.

IOWA—Suzanne Bartlett Hackenmiller

Falzone, Deanna. "The Power of the Great Outdoors and Improved Mental Health." January 28, 2020. https://

www.fox17online.com/rebound/rebound-the-power-of-the-great-outdoors-and-improved-mental-health.

Hackenmiller, Suzanne Bartlett. *The Outdoor Adventurer's Guide to Forest Bathing: Using Shinrin-Yoku to Hike, Bike, Paddles, and Climb Your Way to Health and Happiness.* (Lanham, MD: Falcon Guides, 2019).

———. "On This Earth Day, What Plan Could Revive Our Economy and Heal the Earth?" *Des Moines Register.* April 22, 2020. https://www.desmoinesregister.com/story/opinion/columnists/iowa-view/2020/04/22/green-stimulus-offers-path-forward-covid-19-earth-day/2998001001/.

———. "Finding Healing in Nature Is Important and It Doesn't Require an Ambitious Trip." *Des Moines Register.* November 29, 2020. https://www.desmoinesregister.com/story/opinion/columnists/iowa-view/2020/11/29/finding-healing-nature-doesnt-require-ambitious-journey/6420333002/.

———. "Solving the Climate Crisis: A Trail for Healing for a Planet and Its People." Johns Hopkins Celebrates Earth Day 50. YouTube. January 6, 2021. https://www.youtube.com/watch?v=3tRU9-xnkfU.

KANSAS—Pooja Shaw

Heinrich, Martin. "Your Next Car and Clothes Dryer Could Help Save Our Planet." *New York Times.* June 8, 2021. https://www.nytimes.com/2021/06/08/opinion/climate-change-electricity-fossil-fuels.html.

Shah, Pooja. LinkedIn. 2021. https://www.linkedin.com/in/poojashah157/.

Women in Green Hydrogen. "Pooja Shah." 2021. https://women-in-green-hydrogen.net/pooja-shah/.

Women of Renewable Industries and Sustainable Energy. 2021. https://wrisenergy.org/.

KENTUCKY—Tiffany Bellfield-El-Amin

Bellfield, Tiffany. "The Struggle in Rural Kentucky, Being a Black Producer, Is Real." *New Dream*, May 10, 2019. https://newdream.org/blog/interview-with-tiffany-bellfield-owner-of-ballew-estates.

Crouch, Claire. "Reopening under New Ownership: Couple Revising Lexington's Alfalfa Restaurant." June 15, 2020. *Lex 18*. https://www.lex18.com/rebound/reopening-under-new-ownership-couple-reviving-lexingtons-alfalfa-restaurant.

Eadens, Savannah. "Less than 2 percent of Kentucky Farms Are Black-owned. This Company Is Trying to Change That." *Louisville Courier-Journal*, July 9, 2020. https://www.courier-journal.com/story/life/2020/07/09/agrotourism-company-black-soil-connecting-black-kentucky-farmers/3210436001/.

Herald-Leader, Lexington. "Meet the New Owners of Alfalfa's Restaurant." May 30, 2020. https://www.youtube.com/watch?v=b_3-aym8J_o.

Philpott, Tom. "White People Own 98 percent of Rural Land. Young Black Farmers Want to Reclaim Their Share." *Mother Jones*, June 27, 2020. https://www.motherjones.com/food/2020/06/black-farmers-soul-fire-farm-reparations-african-legacy-agriculture/.

LOUISIANA—Colette Pichon Battle

Battle, Colette Pichon. "Climate Change Will Displace Millions. Here's How We Prepare." TED Talk. 2020. https://

www.ted.com/talks/colette_pichon_battle_climate_
change_will_displace_millions_here_s_how_we_
prepare?language=en.

———. "An Offering from the Bayou." In *All We Can Save:
Truth, Courage, and Solutions for the Climate Crisis*, edited
by Ayana Elizabeth Johnson and Katharine Wilkinson
(New York: One World, 2020), 329–333.

———. "How Can We Prepare for the Next Hurricane Katrina?"
TED Radio Hour. February 26, 2021. https://www.npr.
org/2021/02/26/971498925/colette-pichon-battle-how-
can-we-prepare-for-the-next-hurricane-katrina.

Gulf Coast Center for Law and Policy. "Meet the GCCLP
Krewe: Colette Pichon Battle, Esq." 2021. https://
www.gcclp.org/krewe.

Obama Foundation. "Colette Pichon Battle: Organizing for Black
and Native Climate Justice and Collaboration in the Gulf
South." 2019. Obama Fellowship. https://www.obama.
org/fellowship/2019-fellows/colette-pichon-battle/.

Rein, Marcy, and Jess Clarke. "As the South Goes: Organizing,
Healing, and Resilience in Gulf Coast Communities."
Reimagine Radio. August 2014. https://www.reimag-
inerpe.org/20-1/pichon-battle.

MAINE—Cassie Cain

350 Maine. "Climate Crisis in Maine." 2021. http://web.archive.
org/web/20210620161503/https://www.350maine.
org/climate_change.

Thill, David. "Meet the Young People Pushing Maine Forward on
Climate Change." *Energy News Network*. August 5, 2020.
https://energynews.us/2020/08/05/meet-the-young-
people-pushing-maine-forward-on-climate-change/.

Yared, Katie. "The Youth-led Fight for Climate Justice in Maine." *Climate Xchange*. April 16, 2021. https://climate-xchange.org/2021/04/16/the-youth-led-fight-for-climate-justice-in-maine/.

MARYLAND—Jacqui Patterson

Jacobs, Jason. "Ep 95–Jacqueline Patterson, Senior Director of Environmental and Climate Justice Program at the NAACP." *My Climate Journey*. April 2, 2020. https://podcasts.apple.com/us/podcast/ep-95-jacqueline-patterson-senior-director-environmental/id1462776122?i=1000470284067.

Patterson, Jacqui. "Jacqui Patterson of NAACP at Inspire Speakers Series." September 22, 2020. https://www.youtube.com/watch?v=-Rrvr7padIg.

———. "The Fight Against Climate Racism with NAACP Environmental Justice Director Jacqui Patterson." August 19, 2020. https://www.youtube.com/watch?v=eac5snGnJjE&t=578s.

———. "Environmental Injustices and Climate Change through a Civil Rights Lens." Middlebury Institute of International Studies. October 15, 2020. https://www.youtube.com/watch?v=d01Duj-mMAo.

MASSACHUSETTS—Varshini Prakash

Klein, Ezra. "No Permanent Friends, No Permanent Enemies: Inside the Sunrise Movement's Plan to Save Humanity." *Vox*. July 31, 2019. https://www.vox.com/ezra-klein-show-podcast/2019/7/31/20732041/varshini-prakash-sunrise-movement-green-new-deal.

Prakash, Varshini. "Varshini Prakash on Redefining What's Possible." December 22, 2020. https://www.sierraclub.org/sierra/2021-1-january-february/feature/varshini-prakash-redefining-whats-possible.

Raz, Guy. "How I Built This: Resilience." Varshini Prakash. October 28, 2020. https://www.facebook.com/watch/live/?v=378697396666522&ref=watch_permalink.

MICHIGAN—Mari Copeny

Copeny, Mari. "Mari Copeny." 2021. https://www.maricopeny.com/.

Denchak, Melissa. "Flint Water Crisis. Everything You Need to Know." Natural Resources Defense Council. November 8, 2018. https://www.nrdc.org/stories/flint-water-crisis-everything-you-need-know.

Euse, Erica. "Greta Thunberg Is Not the Only Young Climate Activist You Need to Know." *i-d.Vice*. September 25, 2019. https://i-d.vice.com/en_uk/article/pa7zbb/greta-thunberg-isnt-the-only-young-climate-activist-you-need-to-know.

MINNESOTA—Tara Houska

Houska, Tara. "The Standing Rock Resistance and Our Fight for Indigenous Rights." TED Talk. 2017. https://www.ted.com/talks/tara_houska_the_standing_rock_resistance_and_our_fight_for_indigenous_rights?language=en.

———. "A Voice from the Forest in the Corporate Boardroom." *Aljazeera*. January 1, 2020. https://www.common-dreams.org/views/2020/01/01/voice-forest-corporate-boardroom.

———. "What the Climate Movement Can Learn from Indigenous Values." *Vogue*. September 24, 2020. https://www.vogue.com/article/what-the-climate-movement-can-learn-from-indigenous-values.

———. "Sacred Resistance." In *All We Can Save: Truth, Courage, and Solutions for the Climate Crisis*, edited by Ayana Elizabeth Johnson and Katharine Wilkinson (New York: One World, 2020), 203–219.

Marsh, Steve. "Tara Houska on Racist Mascots, Fighting Pipelines, and Being the Only Native Person in the Room." *Minneapolis-St. Paul*. Oct. 18, 2020.

MISSISSIPPI—Heather McTeer Toney

Mock, Brentin. "Tough Love: Can a Local Leader Save the EPA's Troubled Southeast Region?" *Grist*. January 22, 2014. https://grist.org/climate-energy/tough-love-can-a-local-leader-save-the-epas-troubled-southeast-region/.

Toney, Heather McTeer. "Black Women are Leaders in the Climate Movement." *New York Times*. June 25, 2019. https://www.nytimes.com/2019/07/25/opinion/black-women-leaders-climate-movement.html.

———. "Moving Past Stereotypes: Climate Action IS the Social Justice Issue of Our Time." Keynote to Bioneers Conference. November 13, 2019. https://bioneers.org/moving-past-stereotypes-climate-action-social-justice-heather-mcteer-toney-zstf1911/.

———. "Heather McTeer Toney Speaks at Fire Drill Friday." February 28, 2020. https://www.youtube.com/watch?v=Pz_dUl6wBF0.

———. "Heather McTeer Toney: Rachel's Network Catalyst Award Winner." June 18, 2020. https://www.youtube.com/watch?v=-bR_IqFni1Q.

———. "Collards Are as Good as Kale." In *All We Can Save: Truth, Courage, and Solutions for the Climate Crisis*, edited by Ayana Elizabeth Johnson and Katharine Wilkinson (New York: One World, 2020), 75–83.

MISSOURI—Emily Lecuyer

AltCap. "Impact Investing." 2021. https://www.alt-cap.org/impact-investing.

Feist, Kathy. "Abandoned Marlborough School Gets Investor(s) and New Hope." *Martin City Telegraph.* June 27, 2021. https://martincitytelegraph.com/2021/06/27/abandoned-marlborough-school-gets-investors-and-new-hope/.

"Investing in Impact: Lauren Conaway Talks with Emily Lecuyer." *Startup Hustle: A Podcast for Entrepreneurs.* September 17, 2020. https://podcasts.apple.com/us/podcast/investing-in-impact/id1323580239?i=1000491554632.

MONTANA—Grace Gibson-Snyder

"Held v. State of Montana. Online Briefing." February 11, 2021. https://vimeo.com/511436891.

"Held v. State of Montana et al. Complaint." March 13, 2020. file:///Users/wwc/Downloads/Held%20et%20al.%20v.%20State%20of%20Montana%20et%20al.%20Complaint.pdf.

Pesce, Nicole Lyn. "Young Adults Worry It's 'Morally Wrong' to Have Children." *MarketWatch*. April 24, 2021. https://www.marketwatch.com/story/young-adults-worry-its-morally-wrong-to-have-children-earth-day-study-finds-11619110785.

Our Children's Trust. "Youth v. Gov Montana." 2021. https://www.ourchildrenstrust.org/montana.

NEBRASKA—Jess Benjamin

Benjamin, Jess. "Jess Benjamin Ceramic Artist." 2021. http://www.jessbenjamin.com.

Carpenter, Kim. "Very Serious Play: A Conversation with Jess Benjamin." *Sculpture Magazine*. January 10, 2017. https://sculpturemagazine.art/very-serious-play-a-conversation-with-jess-benjamin/.

Konen, Judee. "Hastings College Grad Explores Art with an Environmental Message." April 12, 2019. https://www.hastings.edu/success-stories/hastings-college-grad-explores-art-with-an-environmental-message/.

NEW MEXICO—Kendra Pinto

Canalos, Asha. "This Is Serious Stuff. This Is Life—An Interview with Kendra Pinto." November 11, 2016. http://asha-canalos.squarespace.com/kendra-pinto.

Hyman, Randall. "New Fossil Fuel Projects Meet Indigenous Resistance in New Mexico." *Truthout*. September 16, 2020. https://truthout.org/articles/new-fossil-fuel-projects-meet-indigenous-resistance-in-new-mexico/.

Pinto, Kendra. "How Are Our Lives Not Important? An Eye-witness Describes the Nageezi, New Mexico Oil Fire." *Huffington Post*. July 22, 2016. https://www.huffpost.com/entry/how-are-our-lives-not-important-an-eyewitness-describes_b_579251cde4b0a1917a6e8725.

———. "Navajo Kendra Pinto Testifies before Congress about the Methane Rule." June 21, 2017. https://www.youtube.com/watch?v=2hlW8uDH4WI.

Reggero, Dayna, and Moms Clean Air Force. "Fortitude: The Power of Past and Present." *The Story We Want*. Nov. 3, 2017. Climate Listening Project. http://climatelistening-project.org/2017/11/04/the-story-we-want-fortitude/.

Wranger, Valory, Herbert Benally, and David Tsosie. *A Cultural, Spiritual, and Health Impact Assessment of Oil and Drilling Operations in the Navajo area of Counselor Chapter, New Mexico*. 2018. https://eplanning.blm.gov/public_projects/nepa/116415/20007002/250008124/NM_20181031_Counselor_Health_Impact_Assessment_Committee.pdf.

NEW YORK—Ayana Elizabeth Johnson

"Billie Eilish Talking about Climate Change with Dr. Ayana Elizabeth Johnson." April 21, 2021. https://www.youtube.com/watch?v=kVjrDyun5ck.

Gardinar, Beth. "Ocean Justice. Where Social Equity and the Climate Fight Intersect." *Yale Environment 360*. July 16, 2020. https://e360.yale.edu/features/ocean-justice-where-social-equity-and-the-climate-fight-intersect.

Johnson, Ayana Elizabeth, and Katharine Wilkinson, eds. *All We Can Save: Truth, Courage, and Solutions for the Climate Crisis*. (New York: One World, 2020).

Johnson, Ayana Elizabeth. "Our Oceans Brim with Climate Solutions. We Need a Blue New Deal." *Washington Post*.

December 10, 2019. https://www.washingtonpost.com/opinions/2019/12/10/green-new-deal-has-big-blue-gap-we-need-protect-our-oceans/.

———. "I'm a Black Climate Scientist. Racism Derails Our Efforts to Save the Planet. *Washington Post*. June 3, 2020. https://www.washingtonpost.com/outlook/2020/06/03/im-black-climate-scientist-racism-derails-our-efforts-save-planet/.

———. How to Find Joy in Climate Action. TED Talk 2022. https://www.ted.com/talks/ayana_elizabeth_johnson_how_to_find_joy_in_climate_action.

Riley-Adams, Ella. "A New Climate Podcast Asks 'Are We Screwed?'—but Still Manages to Be Constructive." *Vogue*. September 18, 2020. https://www.vogue.com/article/ayana-elizabeth-johnson-new-climate-podcast.

Tsui, Bonnie. "Ayana Elizabeth Johnson Is the Climate Leader We Need." *Outside*. October 31, 2020. https://www.outsideonline.com/outdoor-adventure/environment/ayana-elizabeth-johnson-climate-change-leader/.

NEVADA—Cecelia González

Bahouth, Brian. "30 by 30 Joint Resolution Sparks Questions, Highlights Diversity of Environmental Values, Partisan Divides." *Sierra Nevada Daily*. March 11, 2021. https://www.sierranevadaally.org/2021/03/11/30-by-30-joint-resolution-sparks-questions-highlights-diversity-of-environmental-values-partisan-divide/.

Rindels, Michelle. "Freshman Orientation: Assemblywoman Cecelia González." *The Nevada Independent*. February 3, 2021. https://thenevadaindependent.com/article/freshman-orientation-assemblywoman-cecelia-gonzalez.

NEW HAMPSHIRE—Elizabeth Burakowski

Burakowski, Elizabeth. "If you Like Skiing or Snowboarding, You Should Care about Climate Change." *The Guardian*. February 16, 2015. https://www.theguardian.com/commentisfree/2015/feb/16/skiing-snowboarding-winter-snow-climate-change.

———. "Climate, Snow, Society." 2021. https://elizabethburakowski.com/.

———. "Climate Change and the Ski Industry." TEDx Belmont. July 3, 2018. https://www.youtube.com/watch?v=HnTuNzaNqvM.

Flint, Jennifer McFarland. "Hope in Action: A New Climate for Environmental Activism." *Wellesley Magazine*. Fall 2017. https://magazine.wellesley.edu/fall-2017/hope-action.

Ropeik, Annie. "UNH Climate Research: Less Snow Hurts Economies and Environment." *New Hampshire Public Radio*. February 27, 2018. https://www.nhpr.org/environment/2018-02-27/unh-climate-research-less-snow-hurts-economies-environment.

———. "As Winters Warm, New Englanders Are Finding Normal Cold Weather More Unusual." *New Hampshire Public Radio*. March 3, 2019. https://www.nhpr.org/environment/2019-03-04/as-winters-warm-new-englanders-are-finding-normal-cold-weather-more-unusual.

NEW JERSEY—Saarah Yasmin Latif

Majeed, Kori and Saarah Yasmin Latif. *Forty Green Hadith: Sayings of the Prophet Muhammad (PBUH) on Environmental Justice and Sustainability*. tiny.cc/40greenhadith.

The BTS Center. "Saarah Yasmin Latif." 2021. https://thebtscenter.org/saarah-y-latif/.

NORTH CAROLINA—Dayna Reggero

Reggero, Dayna, and Moms Clean Air Force. "Persistence: The Power of People and Prayer." *Climate Listening Project.* May 16, 2017. http://climatelisteningproject.org/2017/05/16/the-story-we-want-persistence/.

Reggero, Dayna. *Planet Prescription: Mothers and Others for Clean Air. Climate Listening Project.* 2020. https://planetprescription.com/about.

NORTH DAKOTA—Kandi Mossett White

Johnston, Jill, Esther Lim, and Hannah Roh. "Impact of Upstream Oil Extraction and Environmental Public Health: A Review of the Literature." *Sci. Total Environment* March 20, 2019. 187–199. https://pubmed.ncbi.nlm.nih.gov/30537580/.

MADRE. "Indigenous Women Fighting for Climate Justice." September 19, 2019. https://www.youtube.com/watch?v=3IAMYJ5bh54.

Susskind, Yifat. "A Mother's Resolve: Organizing for Local and Global Climate Justice." MADRE. May 11, 2019. https://www.madre.org/press-publications/article/mother%E2%80%99s-resolve-organizing-local-and-global-climate-justice.

White, Kandi. "Ties of Blood." Autumn 2018. *Earth Island Journal.* https://www.earthisland.org/journal/index.php/magazine/entry/ties-of-blood/.

OHIO—Jill Antares Hunkler

Eaton, Sabrina. "Ohio Anti-fracking Activist Joins Greta Thunberg to Decry Fossil Fuel Subsidies at Earth Day Congressional Hearing." *Cleveland.com*. April 22, 2021. https://www.cleveland.com/open/2021/04/ohio-anti-fracking-activist-joins-greta-thunberg-to-decry-fossil-fuel-subsidies-at-earth-day-congressional-hearing.html.

Hergesheimer, Courtney. "Fracking in Belmont County, Ohio." Video dispatch. *Columbus Dispatch*. July 28, 2019. https://stories.usatodaynetwork.com/fracking/.

Hunkler, Jill. "The Role of Fossil Fuel Subsidies in Preventing Action of the Climate Crisis. Written Testimony of Jill Antares Hunkler, Seventh Generation Ohio Valley Resident." April 22, 2021. https://www.govinfo.gov/content/pkg/CHRG-117hhrg44383/html/CHRG-117hhrg44383.htm.

Hunkler, Jill and John Stolz. "Establishing a Low-cost Sustainable Environmental Monitoring Program—Barnesville, Ohio." *Thriving Earth Exchange*. 2021. https://thrivingearthexchange.org/project/barnesville-oh/.

OKLAHOMA—Casey Camp-Horinek

Camp-Horinek, Casey. "Aligning Human Law with Natural Law." Bioneers Conference. 2019. https://bioneers.org/casey-camp-horinek-aligning-human-law-natural-law-zstf1911/.

———. "Mother Earth Is Calling Us: A Message from Casey Camp-Horinek for 2021." December 30, 2020. https://www.youtube.com/watch?v=GmIrHt7kt9E.

Camp-Horinek, Casey, and Christine Hanna. "The Work of Breaking Free." *Yes! Magazine.* May 10, 2021. https://www.yesmagazine.org/authors/casey-camp-horinek.

Global Alliance for the Rights of Nature. "Learn about Rights of Nature." 2021. https://www.therightsofnature.org/.

Hutner, Heidi. "Activist Casey Camp-Horinek on Mothering and the Standing Rock Protest." Moms Clean Air Force. October 11, 2016. https://www.momscleanairforce.org/interview-casey-camp-horinek/.

OREGON—Kelsey Juliana

Americans Who Tell the Truth. "Kelsey Juliana." (n.d.) https://www.americanswhotellthetruth.org/portraits/kelsey-juliana. Accessed September 3, 2021.

Our Children's Trust. "Juliana v. US" 2021. https://www.ourchildrenstrust.org/juliana-v-us.

"The Juliana Plaintiffs: Vic Barrett, Levi Draheim, and Kelsey Juliana with Sandra Upson." *WIRED.* November 8, 2019. https://www.facebook.com/watch/live/?v=2525473537776514&ref=watch_permalink.

PENNSYLVANIA—Mollie Michel

Energy Foundation. "Pennsylvania Moms Demand a Clean Energy Future." 2021. https://www.ef.org/story/pennsylvania-moms-demand-a-clean-energy-future/.

Environmental Defense Fund. "EDF Analysis Finds Pennsylvania Oil and Gas Methane Emissions Are Double Previous Estimation." May 13, 2020. https://www.edf

.org/media/edf-analysis-finds-pennsylvania-oil-and-gas-methane-emissions-are-double-previous-estimate.

Moms Clean Air Force. "Pennsylvania State Senator Katie Muth: Stay In and Speak Out for Climate Action." October 15, 2020. https://www.youtube.com/watch?v=nfevS6KbbV0.

RHODE ISLAND—Emily Diamond

Diamond, Emily (Pechar). "Depolarizing Environmental Policy: Identities and Public Opinion on the Environment." PhD dissertation. Duke University Libraries. 2019. https://dukespace.lib.duke.edu/dspace/handle/10161/18686.

———. "Personal Identity and Climate Change." TEDx URI. April 2021. https://www.ted.com/talks/emily_diamond_personal_identity_and_climate_change.

Mufson, Steven, Chris Mooney, Juliet Eilperin, and John Muyskens. "America's Hot Spots: Rhode Island among the Fastest Warming States in the US." *Providence Journal.* August 14, 2019. https://www.providencejournal.com/news/20190814/americas-hot-spots-ri-among-fastest-warming-states-in-us.

SOUTH CAROLINA—Latria Graham

Dartmouth Sustainability. "Environmental Changemakers: Latria Graham '08." January 15, 2021. https://www.youtube.com/watch?v=_XZWLHY7FAk.

Graham, Latria. "We're Here. You Just Don't See Us." *Outside.* May 1, 2018. https://www.outsideonline.com/culture/opinion/were-here-you-just-dont-see-us/.

———. "A Dream Uprooted." *Garden & Gun.* April/May 2020. https://gardenandgun.com/articles/a-dream-uprooted/.

———. "Out There, Nobody Can Hear You Scream." *Outside Magazine.* September 21, 2020. https://www.outside-online.com/culture/essays-culture/out-there-nobody-can-hear-you-scream/.

———. "Latria Graham: Social Issues Deserve Subplots." 2021. https://www.latriagraham.com/.

Longform Podcast. "Latria Graham. #413." October 9, 2020. https://longform.org/player/longform-podcast-413-latria-graham.

SOUTH DAKOTA—Jasilyn Charger

Charger, Jasilyn. "Jasilyn Charger." In *How We Go Home: Voices from Indigenous North America*, edited by Sara Sinclair. (Chicago: Haymarket Books, 2020), 33–54.

Elbein, Saul. "The Youth Group That Launched a Movement at Standing Rock." *New York Times Magazine.* January 31, 2017. https://www.nytimes.com/2017/01/31/magazine/the-youth-group-that-launched-a-movement-at-standing-rock.html.

Lakota People's Law Project. "Two Native Americans Arrested over Keystone XL Protests." *Indian Country Today.* January 8, 2021. https://indiancountrytoday.com/the-press-pool/two-native-americans-arrested-over-keystone-xl-protests.

Levi's. "Jasilyn: Activist of the Land." March 7, 2018. https://www.facebook.com/watch/?v=101554 56891415662.

TENNESSEE—Sarah Bellos

Bellos, Sarah. "Natural Indigo Dyeing Iron Vat How-to Video." July 13, 2020. https://www.youtube.com/watch?v=1cb-nMtbhBPI.

ChemSec Marketplace. "High Purity US Grown Natural Indigo." n.d. Accessed September 3, 2021. https://market place.chemsec.org/Alternative/High-purity-US-grown-natural-indigo–certified-as-bio-based-renew-able-indigo--215.

Cotton & Moss. "Stony Creek Colors." February 1, 2019. https://cottonandmoss.com/blogs/farm-fiber-sustainability/stony-creek-colors.

Henderson, Nancy. "How Stony Creek Colors Works with Tobacco Farmers to Grow Indigo." *Tennessee Home & Farm*. August 13, 2017. https://www.tnhomeandfarm.com/agriculture/stony-creek-colors-works-tn-farmers-grow-indigo/.

Stony Creek Colors. "Color with life®." 2021. https://stony-creekcolors.com/.

Velasquez, Angela. "Natural Indigo Maker Stony Creek Colors Raises $9 Million on Series B." *Rivet: Sourcing Journal*. March 2, 2021. https://sourcingjournal.com/denim/denim-business/stony-creek-colors-natural-indigo-dye-series-b-sarah-bellos-265180/.

TEXAS—Katharine Hayhoe

Goldberg, M., A. Gustafson, S. Rosenthal, J. Kotcher, E. Maibach, and A. Leiserowitz. *For the First Time, the Alarmed are Now the Largest of the Global Warming's Six Americas*. Yale University and George Mason University. (New

Haven, CT: Yale Program on Climate Communications, 2020).

Hayhoe, Katharine. "The Most Important Thing You Can Do to Fight Climate Change: Talk about It." TED Talk. 2019. https://www.youtube.com/watch?v=-BvcToPZCLI.

———. "I'm a Climate Scientist Who Believes in God. Hear Me Out." *New York Times*. October 31, 2019. https://www.nytimes.com/2019/10/31/opinion/sunday/climate-change-evangelical-christian.html.

———. *Saving Us: A Climate Scientist's Case for Hope and Healing in a Divided World*. (New York: Atria/One Signal, 2021).

———. "Katharine Hayhoe: Climate Scientist." 2021. http://www.katharinehayhoe.com/.

A. Leiserowitz, E. Miabach, S. Rosenthal, J. Kotcher, J, Caroman, X. Wang, J. Marlan, K. Lacroix, and M. Goldberg. *Climate Change in the American Mind*. (New Haven, CT: Yale Program on Climate Change Communication, 2021).

Shepherd, Marshall. "Three Reasons Scientists Endure Social Media Trolls and Attacks." *Forbes*. December 6, 2020. https://www.forbes.com/sites/marshallshepherd/2020/12/06/3-reasons-scientists-endure-social-media-trolls-and-attacks/?sh=6f32bd64424c.

UTAH—Terry Tempest Williams

Fredericksen, Devon. "Ground-truthing." *Guernica*. August 1, 2013. https://www.guernicamag.com/ground-truthing/.

Gay, Roxane. "The Rumpus Interview with Terry Tempest Williams." *The Rumpus*. July 4, 2013. https://therumpus.net/2013/07/the-rumpus-interview-with-terry-tempest-williams/.

Williams, Terry Tempest. *Refuge: An Unnatural History of Family and Place.* (New York: Vintage Books, 1991).

———. *When Women Were Birds: Fifty-four Variations on Voice.* (New York: Picador, 2012.)

———. "Keeping My Fossil Fuel in the Ground." March 29, 2016. *New York Times.* https://www.nytimes.com/2016/03/29/opinion/keeping-my-fossil-fuel-in-the-ground.html.

van Gelder, Sarah. "Terry Tempest Williams: 'Survival Becomes a Spiritual Practice.'" *YES! Magazine.* October 5, 2015. https://www.yesmagazine.org/issue/debt/2015/10/05/terry-tempest-williams-survival-becomes-a-spiritual-practice.

VERMONT—Devi Lockwood

Lockwood, Devi. "Learning to Scale Peaks from my Underprotective Mother." *New York Times.* July 22, 2016. https://well.blogs.nytimes.com/2016/07/22/learning-to-scale-peaks-from-my-underprotective-mother/.

———. "Meet the Globe-trotting Cyclist Collecting 1,001 Climate Change Stories." *The Guardian.* April 28, 2017. https://www.theguardian.com/us-news/2017/apr/28/devi-lockwood-climate-march-1001-climate-change-stories.

———. *1001 Voices on Climate Change: Everyday Stories of Flood, Fire, Drought, and Displacement from Around the World.* (New York: Simon & Schuster, 2021).

———. "Devi Lockwood." 2021. http://devi-lockwood.com.

VIRGINIA—Sophia Kianni

Ashoka Changemakers. "Sophia's Journey to Making the Climate Movement More Accessible." August 25, 2020.

https://www.ashoka.org/en/story/sophia-k-change-maker-story.

Climate Cardinals. "Welcome to Climate Cardinals." 2021. https://www.climatecardinals.org/.

CNN. "How the Pandemic Is Shaping This Globally-minded Teen's Future." *Voices of the Pandemic.* July 2, 2020. https://www.cnn.com/videos/tv/2020/07/02/coronavirus-pandemic-virginia-youth-climate-change-activism-spc.cnn.

Felton, Lena. "Meet the 17-year-old Climate Activist Who Skipped School to Hunger Strike at the Capital." *The Lilly.* November 18, 2019. https://www.thelily.com/meet-the-17-year-old-climate-activist-who-skipped-school-to-hunger-strike-at-the-capitol/.

Ferdowski, Samir. "The Activist Translating Climate Crisis Information across the Globe." *Vice.* December 18, 2020. https://www.vice.com/en/article/7k94dd/the-activist-translating-climate-crisis-information-across-the-globe.

Kart, Jeff. "Youth Activist Uses Quarantine to Start Nonprofit That Translates Climate Change Information from English to Other Languages." *Forbes.* May 12, 2020. https://tinyurl.com/yfzuph6n.

Kianni, Sophia. "What It's Really like to Be a Climate Change Activist in Quarantine?" *Refinery 29.* April 22, 2020. https://www.refinery29.com/en-us/2020/04/9700125/climate-change-activist-diary-sophia-kianni.

WASHINGTON—Jamie Margolin

Art Majors. "Art Majors 2021| Official Pilot Trailer| LGBTQ+ Web Series." June 17, 2021. https://www.youtube.com/watch?v=XoKlxmCDZ18.

Greenpeace. "Fireside Fire Drill Friday with Jane Fonda and Jamie Margolin." June 26, 2020. https://www.facebook.com/watch/live/?v=2689715924642711&ref=watch_permalink.

Jarvis, Brooke. "The Teenagers at the End of the World." *New York Times*. July 21, 2020. https://www.nytimes.com/interactive/2020/07/21/magazine/teenage–activist–climate-change.html.

Margolin, Jamie. *Youth to Power: Your Voice and How to Use It.* (New York: Hachette, 2020).

Soul Pancake. "So You Want to Fight Climate Change? Now What?" November 16, 2020. https://www.youtube.com/watch?v=pCYU9qCjrf8.

WEST VIRGINIA—Mary Anne Hitt

Grist staff. "Mary Anne Hitt, Director of Appalachian Voices, Answers Questions." *Grist*. May 27, 2007. https://grist.org/article/hitt/.

Hitt, Mary Anne. "Beyond Coal." In *All We Can Save: Truth, Courage, and Solutions for the Climate Crisis*, edited by Ayana Elizabeth Johnson and Katharine Wilkinson (New York: One World, 2020), 61–73.

Matthews, Amy. "SkyTruth Spotlight: SkyTruth Board Member Mary Anne Hitt: Activist Extraordinaire." June 3, 2020. https://skytruth.org/2020/06/skytruth-board-member-mary-anne-hitt-activist-extraordinaire/.

WISCONSIN—Tia Nelson

Citizen's Climate Lobby. "Citizen's Climate Lobby monthly meeting with Tia Nelson." Nov. 14, 2020. https://www.youtube.com/watch?v=iqHasAvYNTg.

Nelson, Tia. "The Past and the President: A Commitment to International Climate Action." *Medium*. April 19, 2021. https://tianelson.medium.com/.

Opoien, Jessie. "Tia Nelson Steps Down from Public Lands Board That Banned Climate Change Discussion." *The Capital Times*. July 21, 2015. https://tinyurl.com/mrys7h7b.

Outrider Foundation. "When the Earth Moves." April 15, 2020. https://www.youtube.com/watch?v=Cwnkw5pnG38.

WisPolitics.com. "Interview with Tia Nelson." April 20, 2020. https://www.wispolitics.com/2020/wisopinion-com-earth-day-interview-with-the-outrider-foundations-tia-nelson/.

WYOMING—Merav Ben-David

Ben-David, Merav. "Merav Ben-David for US Senate." 2020. https://www.bendavid2020.com/.

Cahan, Eli. "In Wyoming, an Ecologist Seeks a New Niche as a US Senator." *Science*. October 19, 2020. https://www.science.org/news/2020/10/wyoming-ecologist-seeks-new-niche-us-senator.

TOI Staff. "Israeli-born Merav Ben-David Fails in Wyoming Senate Bid." *The Times of Israel*. November 4, 2020. https://www.timesofisrael.com/israeli-born-merav-ben-david-fails-in-wyoming-senate-bid/.

Looking Forward: This Love Story Isn't Over Yet

Hayhoe, Katharine. *Saving Us: A Climate Scientist's Case for Hope and Healing in a Divided World*. (New York: Atria/One Signal, 2021).

BIBLIOGRAPHY

Johnson, Ayana Elizabeth and Katharine Wilkinson, eds. *All We Can Save: Truth, Courage, and Solutions for the Climate Crisis*. (New York: One World, 2020).

Menakem, Resma. *My Grandmother's Hands: Racialized Trauma and the Pathway to Mending Our Hearts and Our Bodies*. (Las Vegas, NV: Central Recovery Press, 2017).

Solnit, Rebecca. "Ten Ways to Confront the Climate Crisis without Losing Hope." *The Guardian*. November 18, 2021. https://www.theguardian.com/environment/2021/nov/18/ten-ways-confront-climate-crisis-without-losing-hope-rebecca-solnit-reconstruction-after-covid.

Thomas, Leah. Instagram post. "Black Climate Optimism." May 25, 2021.

Westervelt, Amy. "The Cult of Personal Responsibility Is Killing Us." August 17, 2021. https://the-contract.ghost.io/the-cult-of-personal-responsibility-is-killing-us/.

Wilkinson, Katharine. "The Climate Crisis Is a Call to Action. These 5 Steps Helped Me Figure Out How to Be of Use." *Time*. June 19, 2021. https://time.com/6071765/what-can-i-do-to-fight-climate-change/.

ABOUT THE AUTHOR

Mallory McDuff

Author photo by Gary Peeples

Mallory McDuff teaches environmental education at Warren Wilson College in the Swannanoa Valley of Western North Carolina, where students integrate academics, work, and community engagement. With her two daughters, she lives on campus in a 900-square-foot rental with an expansive view of the Appalachian Mountains, a herd of cows, and a stubborn donkey named Tallulah. She grew up in Fairhope, Alabama, exploring the red-clay bluffs and estuarine waters of Mobile Bay. Her writing examines the intersection of people and places for a better world for all.

She's the author of numerous books, most recently *Our Last Best Act: Planning for the End of Our Lives to Protect the People and Places We Love*, as well as *Sacred Acts: How Churches are Working to Protect Earth's Climate*, and *Natural Saints: How People of Faith are Working to Save God's Earth*. She also coauthored *Conservation Education and Outreach Techniques*. Mallory has written more than fifty essays for publications such as *The New York Times*, *The Washington Post*, *WIRED*, and more.

ABOUT THE AUTHOR

She received her PhD from the University of Florida, MA from the University of South Alabama, and BS from Vanderbilt University. Connect with her at mallorymcduff.com, as well as on social media: Facebook: @mallory.mcduff Instagram: @mallorymcduff1 Twitter: @malmcduff

PHOTO CREDITS

Alaska   Bernadette Demientieff (Photo courtesy of: Bernadette Demientieff, used by permission)

Alabama   Anna Jane Joyner (Photo courtesy of: Anna Jane Joyner, used by permission)

Arizona   Diana Liverman (Photo courtesy of: Diana Liverman, used by permission)

Arkansas   Donna Kilpatrick (Photo courtesy of: Phil Davis, used by permission)

California   Amanda Gorman (Photo Credit: National Youth Poet Laureate Amanda Gorman reads her work, "An American Lyric," at the inaugural reading of Poet Laureate Tracy K. Smith, September 13, 2017. https://blogs.loc.gov/loc/2021/01/amanda-gorman-selected-as-president-elect-joe-bidens-inaugural-poet/ (image: Shawn Miller) https://commons.wikimedia.org/wiki/File:Amanda_Gorman_2017_(cropped).jpg)

Colorado   Beth Conover (Photo courtesy of: Beth Conover, used by permission)

Connecticut   Wawa Gatheru (Photo courtesy of: Wanjiku Gatheru, used by permission)

Delaware   Lisa Locke (Photo courtesy of: Lisa Locke, used by permission)

Florida   Caroline Lewis (Photo courtesy of: Caroline Lewis, used by permission)

PHOTO CREDITS

Georgia	Katharine Wilkinson (Photo courtesy of: Ben Brinker, used by permission)
Hawai'i	Mackenzie Feldman (Photo courtesy of: Mackenzie Feldman, used by permission)
Idaho	Jennifer Ladino (Photo courtesy of: Jennifer Ladino, used by permission)
Illinois & DC	Rhiana Gunn-Wright (Photo Credit: Road to a Green New Deal Tour stop feat. AOC & Bernie Sanders! – LiveStream) Used under Creative Commons License.

Rhiana Gunn-Wright, policy lead for the GND, New Consensus. *This screenshot excerpt was originally uploaded on YouTube under a CC license.* "YouTube allows users to mark their videos with a Creative Commons CC BY license. https://commons.wikimedia.org/wiki/File:Rhiana_Gunn-Wright_at_Road_to_a_Green_New_Deal_Tour_final_stop.jpg

Indiana	Lou Weber (Photo courtesy of: Lou Weber, used by permission)
Iowa	Suzanne Bartlett Hackenmiller (Photo courtesy of: Suzanne Bartlett Hackenmiller, used by permission)
Kansas	Pooja Shah (Photo courtesy of: Pooja Shah, used by permission)
Kentucky	Tiffany Bellfield-El-Amin (Photo courtesy of: Tiffany Bellfield-El-Amin, used by permission)
Louisiana	Colette Pichon Battle (Photo courtesy of: TED, used by permission)
Maine	Cassie Cain (Photo courtesy of: Cassie Cain, used by permission)
Massachusetts	Varshini Prakash (Photo Credit: Publisher's Note: Cropped for purposes of the book. © Attribution: The Laura Flanders Show This file is licensed under the Creative Commons

PHOTO CREDITS

	Attribution 3.0 Unported license. See Link, https://creativecommons.org/licenses/by/3.0/deed.en
Maryland	Jacqui Patterson (Photo courtesy of: Jacqui Patterson, used by permission)
Michigan	Mari Copeny (Photo courtesy of Loui Brezzell, used by permission)
Minnesota	Tara Houska
	Photo Credit: Publisher's Note: Slight photo cropping for purposes of the book. This file is licensed under the Creative Commons Attribution–Share Alike 4.0 International license. See Link: https://creativecommons.org/licenses/by-sa/4.0/deed.en
Mississippi	Heather McTeer Toney (Photo courtesy of: Heather McTeer Toney, used by permission)
Missouri	Emily Lecuyer (Photo courtesy of: Emily Lecuyer, used by permission)
Montana	Grace Gibson-Snyder (Photo courtesy of: Grace Gibson-Snyder, used by permission)
Nebraska	Jess Benjamin (Photo courtesy of: Jess Benjamin, used by permission)
Nevada	Cecelia González (Photo courtesy of: Cecelia González, used by permission)
New Hampshire	Elizabeth Burakowski (Photo courtesy of: Elizabeth Burakowski, used by permission)
New Jersey	Saarah Yasmin Latif (Photo courtesy of: Salma Latif, used by permission)
New Mexico	Kendra Pinto (Photo courtesy of: Kendra Pinto, used by permission)
New York	Ayana Elizabeth Johnson (Photo courtesy of: Ryan Lash, used by permission)
North Carolina	Dayna Reggero (Photo courtesy of: Loriel, shotbyloriel.com, used by permission)

PHOTO CREDITS

North Dakota Kandi Mossett White (Photo courtesy of: Kandi Mossett White, used by permission)

Ohio Jill Antares Hunkler (Photo courtesy of: Jill Antares Hunkler, used by permission)

Oklahoma Casey Camp-Horink (Photo courtesy of: Casey Camp-Horink, used by permission)

Oregon Kelsey Juliana (Photo courtesy of: Kelsey Juliana, used by permission)

Pennsylvania Mollie Michel (Photo courtesy of: Mollie Michel, used by permission)

Rhode Island Emily Diamond (Photo courtesy of: Emily Diamond, used by permission)

South Carolina Latria Graham (Photo courtesy of: Kim Cross, used by permission)

South Dakota Jasilyn Charger (Photo credit: Dawnee Lebeau)

Tennessee Sarah Bellos (Photo courtesy of: Sarah Bellos, used by permission)

Texas Katharine Hayhoe (Photo courtesy of: Ashley Rodgers, Texas Tech University, used by permission)

Utah Terry Tempest Williams (Photo courtesy of: Terry Tempest Williams, used by permission)

Vermont Devi Lockwood (Photo courtesy of: Devi Lockwood, used by permission)

Virginia Sophia Kianni (Photo courtesy of: Sophia Kianni, used by permission)

Washington Jamie Margolin (Photo courtesy of: Jamie Margolin, used by permission)

West Virginia Mary Anne Hitt (Photo courtesy of: Molly Humphreys, Piccadilly Posh, used by permission)

Wisconsin Tia Nelson (Photo courtesy of: Tia Nelson, used by permission)

Wyoming Merav Ben-David (Photo courtesy of Merav Ben-David, used by permission)